PLASTIC DESIGN OF FRAMES

1

FUNDAMENTALS

PLASTIC DESIGN OF FRAMES

LORD BAKER

Emeritus Professor of Mechanical Sciences
University of Cambridge

AND

JACQUES HEYMAN

Professor of Engineering
University of Cambridge

1. FUNDAMENTALS

CAMBRIDGE UNIVERSITY PRESS

CAMBRIDGE

LONDON   NEW YORK   NEW ROCHELLE
MELBOURNE   SYDNEY

CAMBRIDGE
UNIVERSITY PRESS

32 Avenue of the Americas, New York NY 10013-2473, USA

Cambridge University Press is part of the University of Cambridge.

It furthers the University's mission by disseminating knowledge in the pursuit of
education, learning and research at the highest international levels of excellence.

www.cambridge.org
Information on this title: www.cambridge.org/9780521297783

First published 1969
First paperback edition 1980

A catalogue record for this publication is available from the British Library

Library of Congress Catalogue Card Number: 69–19370

ISBN 978-0-521-07517-6 Hardback
ISBN 978-0-521-29778-3 Paperback

CONTENTS

v

CONTENTS

PREFACE

This book presents the basic ideas of simple plastic theory, and should cover the needs of a student of structural engineering up to the level of a first degree. A first course might use material from chapters 1 and 2 only (perhaps omitting sections 1.6 to 1.8, 2.4 and 2.8) in which direct solutions are obtained to problems of analysis and design. The more sophisticated techniques of chapters 3 and 4 can then be studied, both for their scientific importance, and because they enable a much wider range of structures to be examined.

Each chapter contains a set of examples for the student to work, roughly graded from very easy to really quite difficult. Indeed, the reader who can solve all the examples of chapter 4 may conclude that he has mastered the subject. Since plastic theory makes possible the direct design of steel frames in a way that is not possible with elastic methods, there is some emphasis on practical design problems. The undergraduate may wish to omit most of these problems from a preliminary study of the text, whereas a designer might tackle these first. (The abbreviation $M.S.T.$ II, followed by a date after some of the examples, indicates that they have been taken from the papers set for Part II of the Mechanical Sciences Tripos, Cambridge, for that year.)

There are many topics which have not been discussed in this volume, either because of their complexity or because they are not of primary importance. Notes have been made in the text to indicate where a discussion is incomplete, and volume 2 will give further consideration to these and other applications of the theory.

Dr W. H. Ng has read the text with great care, but the responsibility for any remaining errors must lie with the authors. Miss I. Bowen has typed the manuscript, as she did those for both volumes of *The Steel Skeleton*, with her usual good temper and skill.

1

THE PLASTIC HINGE

1.1 The collapse of beams

An engineering structure has to satisfy many functional requirements; of these, the most important are that it shall be *strong* enough to resist the external loading (and its own weight) without collapsing, and *stiff* enough not to deflect unduly under that loading. Thus if a simply-supported steel roof beam, fig. 1.1, is considered as a typical structural element, then the beam must carry the load W over the span l, and must do so without excessive deflexion.

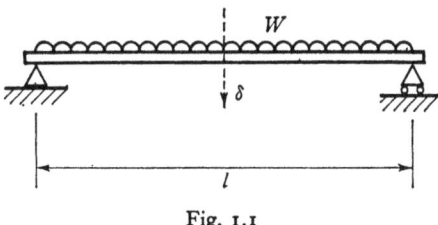

Fig. 1.1

In the conventional engineering practice for building frames, it has been usual to design such structural elements on the basis of their *elastic* behaviour, that is, to attempt to ensure that no permanent deformation occurs under the most unfavourable loading conditions. On this basis, the strength of the structure is assessed by observing or calculating how close the structure is to yelding in any of its parts.

Such an assessment of strength is somewhat arbitrary, and may be examined in the light of the actual behaviour of the simply-supported beam as the load W is slowly increased. If the value of the load is plotted against that of the central deflexion δ, a load-deflexion curve of the general form of fig. 1.2 would be observed. From O to A the behaviour is indeed elastic; the deflexions of the beam are fully recoverable on removal of the load, and no permanent deformation occurs. As the load is further increased, however, there is some permanent set of the beam, and deflexions increase very much more rapidly with increase of loading along the portion AB of the curve. If the load is reduced in this region, unloading of the beam will occur elastically; if the load is removed

I

completely, the beam will be seen to have roughly the shape sketched in fig. 1.3, with a more or less definite 'kink' at the central cross-section.

Further increase of the load W, along the portion BC of the curve in fig. 1.2, leads to a very rapid increase of deflexion. The rising character-istic of the load-deflexion curve is in general due to strain-hardening of the material. The slope of this portion of the curve depends also,

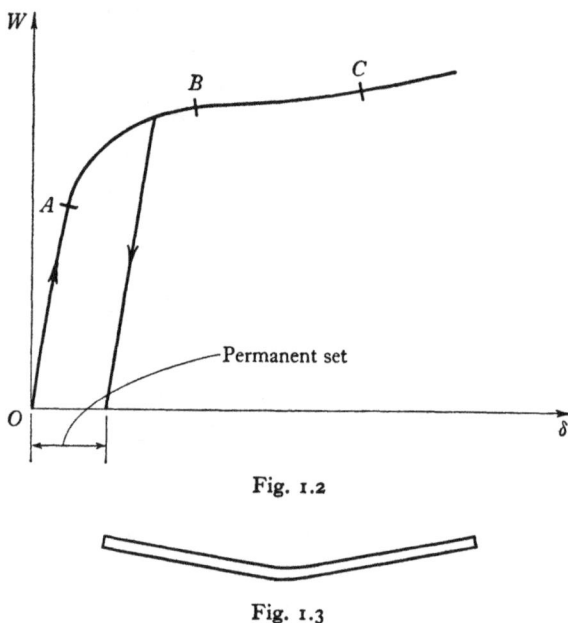

Fig. 1.2

Fig. 1.3

however, on the loading conditions; the sketch is approximately correct for a beam made of structural mild steel (Grade 43 A, BS 4360: 1968),† loaded with a uniformly distributed load, but the curve would be steeper for the same beam subjected to a central point load. In either case deflexions are so large in this region that the beam has probably reached the limit of its useful life, and it is justifiable to consider the beam to have collapsed when the load has reached the value corresponding to B in fig. 1.2.

As an approximation, therefore, the curve of fig. 1.2 may be replaced by the idealized curve of fig. 1.4, for which deflexions increase without

† *British Standard 4360: Weldable Structural Steels* deals with the various grades of steel usually used for general building construction. The Standard lays down certain material specifications, of which the two most important from the point of view of plastic theory are a guaranteed yield stress and a minimum ductility.

limit, at a constant load W_c, along the portion BC of the curve. The load W_c is called the *collapse load* of the beam. Since the collapse load is constant in fig. 1.4, then the bending moments in the collapsing beam are also constant (at least to a first approximation). The unrestricted deflexions at collapse are produced by extremely localized deformation at the central 'kink' in the beam. This central cross-section behaves

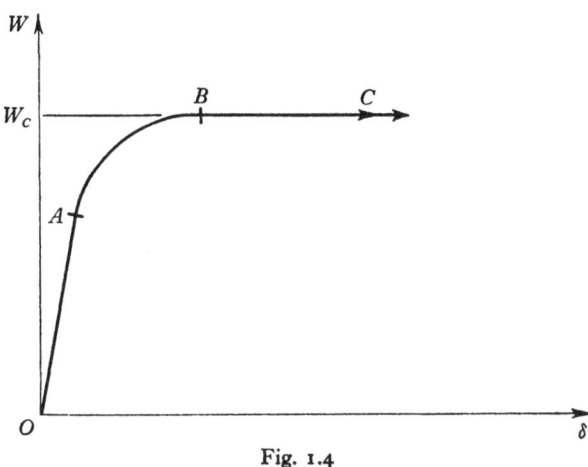

Fig. 1.4

like a 'rusty' hinge connecting together the two halves of the beam, which are, by comparison, almost completely rigid. The full collapse load W_c must act if the hinge is to turn, and collapse occurs as a quasi-static process. Even if deflexions have become quite large, any small reduction δW in the value of the load will ensure that the beam becomes a structure once more, capable of supporting indefinitely without further deflexion the slightly reduced load.

In its collapsed state, the beam has the shape sketched in fig. 1.3, and almost all of the large collapse deflexions of fig. 1.4 may be considered as due to the rotation of the central hinge. The hinge is known as a *plastic hinge*; since it occurs at the section of greatest bending moment in the beam, its formation corresponds to some maximum limit to the bending moment that can be imposed on the beam.

For this simple model, the strength of the structure is given by its collapse load W_c, and is not immediately related to the elastic behaviour, as is conventionally assumed. Although the load at which yield occurs, point A in fig. 1.4, bears some more or less definite ratio to the collapse

3

load for the simply-supported beam, this ratio will vary according to the type of structure considered. The mere onset of yielding does not imply that collapse is imminent.

However, the second structural requirement, that of stiffness, must be considered. It turns out that there is a large class of building structure, of which the multi-storey steel frame is an example, for which the designer is very seldom worried by deflexions. That is, if the members are proportioned on the basis of their strengths, and then calculations are made to estimate deflexions, these second calculations are rarely critical. It is for this class of structure that plastic theory has been developed.

Simple plastic theory, in common with elastic theory and the conventional 'working stress' method of structural design, assumes in the first instance that deflexions of a structure are not the critical design criteria; thus the structure as a whole, and its components, can be proportioned on the basis of the collapse load. Naturally, some margin of safety is incorporated in the design, as will be seen; the structure is not proportioned to collapse under the *working* values of the loads. The designer cannot, of course, be relieved of the responsibility of checking deflexions; if he thinks that these might be large enough to cause inconvenience, then he must make checks in the usual way.

Large deflexions will cause inconvenience to the user, but there is another technical sense in which the effect of deflexions must be taken to be secondary. Both elastic and plastic methods assume that deflexions are, in any case, small compared with the overall dimensions of a frame; that is to say, the geometry of the structure is not sensibly altered by the application of the loads. Thus the deflexion at the points A or B in fig. 1.4 is in reality supposed to be very small compared with the span l of the beam, and the distorted shape of fig. 1.3 has been exaggerated to illustrate the 'kinking' at the plastic hinge.

1.2 The full plastic moment

The load deflexion curve of fig. 1.2 (or the idealized curve of fig. 1.4) results from a certain moment–curvature relationship for the cross-section of the simply-supported steel beam. For this simple example, the moment–curvature relationship, fig. 1.5, is similar in outline to the load-deflexion characteristic for the beam, fig. 1.2. As before, an ideal form (fig. 1.6) may be drawn of fig. 1.5, in which the curvature increases indefinitely at a constant value of the bending moment, M_p. Evidently the value M_p corresponds to the moment acting at the plastic hinge; this moment is known as the *full plastic moment*.

The reason for the adjective *full* will be given below, but it will be appreciated that the portion AB of the curve in fig. 1.6 represents a state

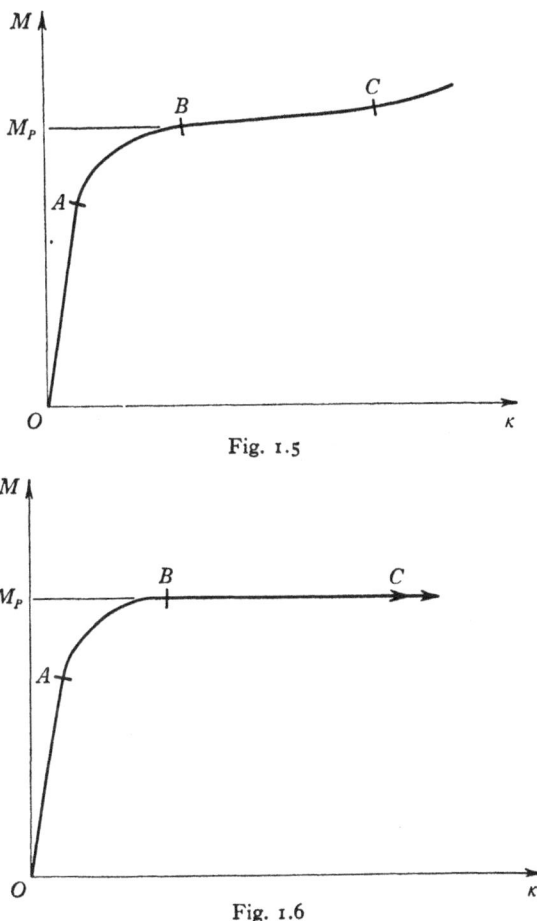

Fig. 1.5

Fig. 1.6

of partial plasticity. From O to A the behaviour is elastic and reversible; a small increase in applied bending moment produces a small proportional change in curvature. If the bending moment is increased above the value corresponding to the point A, then yield will take place, with a corresponding greater increase in curvature. At B the plastic hinge is fully developed, and unlimited increase in curvature, or *rotation*, can occur at the constant bending moment M_p.

5

A knowledge of the value of the full plastic moment is all that is required in simple plastic theory. As will be seen, if the full plastic moments of the various members of a frame are known, then the collapse load of that frame can be determined quite quickly, even if the frame is complex. Similarly, the design of a frame to carry given loads consists in the assignment of certain minimum values of full plastic moment to the members. A simple bend test producing a load-deflexion curve such as that of fig. 1.2 will give an estimate of the value of the full plastic moment, and thus furnish the designer with the information he needs.

However, the strength of structural steel is, as yet, specified in terms of a guaranteed yield stress of the basic material, rather than in terms of a bend test on a structural element, and it is natural to try to relate the fully plastic behaviour in bending to the observed stress–strain relationship in simple tension. The designer would then be able to calculate, or tabulate, sets of *plastic moduli* which, when multiplied by the yield stress of the material he is using, would give the required values of the full plastic moments.

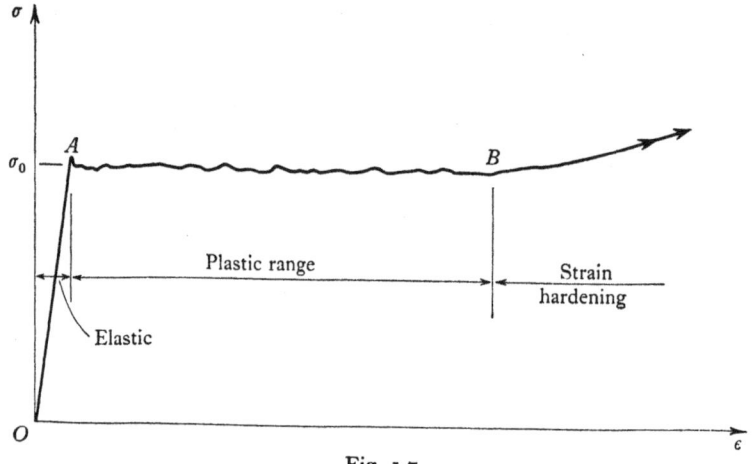

Fig. 1.7

1.3   The bending of beams

The stress–strain curve for a structural mild steel has the general appearance of the curve sketched in fig. 1.7. Elastic extension occurs until the yield stress σ_0 is reached. Thereafter the stress remains sensibly constant until quite large strains have developed, when strain-hardening occurs, and further straining is accompanied by a slow increase in stress. Eventually, at very large strains, necking and fracture will take place.

The strain at first yield, point A in fig. 1.7, is of the order 10^{-3}, i.e. $\frac{1}{10}\%$. The strain at the end of the *plastic range*, i.e. at B, is at least ten times this value for a structural mild steel, lying between 1 and 2%. (The total elongation to fracture might be 50%.) The plastic range is

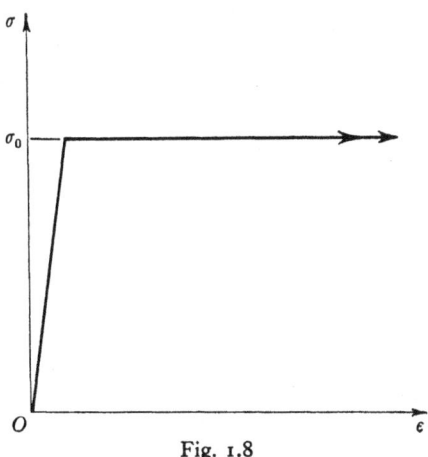

Fig. 1.8

sufficiently long for it to seem reasonable to extend it without limit, that is, to ignore the effect of strain-hardening. Thus in the ideal stress–strain relationship of fig. 1.8, extension is supposed to be unlimited at the constant yield stress σ_0. The error resulting from ignoring the strain-hardening of the material is intuitively on the safe side; a numerical calculation is made below on a beam of rectangular cross-section, which shows that the full plastic moment has been attained for all practical purposes by the time the material starts to strain-harden.

In the work that follows, it will be assumed that behaviour in compression is identical with that in tension. Several of the other assumptions made in the simple elastic theory of bending of beams will also be made in the elastic–plastic and fully plastic analysis. Among these are that strain varies linearly across the cross-section (plane sections remain plane), that adjacent fibres do not affect each other, and that the effect of shearing stress on bending can be ignored. The effect of shear force is in fact discussed more fully below (section 1.7).

Various stages in the bending of a beam of rectangular cross-section, $b \times 2d$, are shown in fig. 1.9. At the onset of yield, the value of the *yield moment* is given by

$$M_y = \tfrac{1}{2}bd\sigma_0(\tfrac{4}{3}d) = \tfrac{2}{3}bd^2\sigma_0. \tag{1.1}$$

The elastic section modulus, Z_e, has value $\frac{2}{3}bd^2$. In the partially plastic state, the value of the moment of resistance of the beam, computed in terms of the parameter α defining the depth of the elastic core, is

$$M_R = (1-\alpha)\,bd\sigma_0[(1+\alpha)d] + \tfrac{2}{3}\alpha^2 bd^2\sigma_0,$$

that is
$$M_R = bd^2\sigma_0(1 - \tfrac{1}{3}\alpha^2). \qquad (1.2)$$

As a check, the value of M_R from equation (1.2) reduces to the value of M_y from equation (1.1) when $\alpha = 1$.

It will be seen that as the value of α approaches zero, that is, as more and more of the cross-section passes from the elastic into the plastic state, the value of the moment of resistance M_R, equation (1.2), approaches the limiting value

$$M_p = bd^2\sigma_0. \qquad (1.3)$$

The *plastic modulus* of a section can be defined by analogy with the elastic modulus; it is that quantity which, when multiplied by the yield stress of the material, will give the value of the full plastic moment M_p. From equation (1.3) it will be seen that the plastic section modulus for the rectangular section is $Z_p = bd^2$. The *shape factor*, defined as the ratio of the plastic to the elastic modulus, thus has the value of $1\cdot5$ for a beam of rectangular cross-section. As will be seen, the value of the shape factor depends only on the geometry of the cross-section, and is about $1\cdot15$ for rolled Universal Beams.

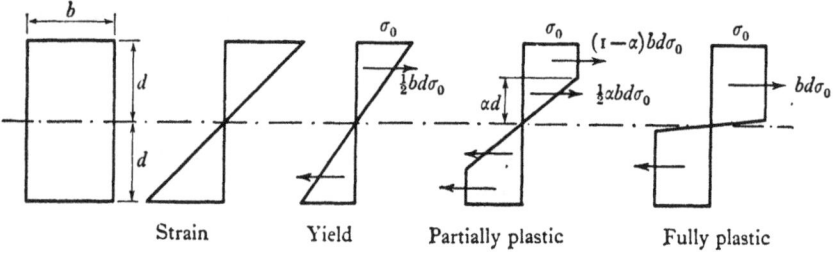

Strain   Yield   Partially plastic   Fully plastic

Fig. 1.9

The state of full plasticity, $\alpha = 0$ in equation (1.2), implies infinite strains in the cross-section; it will be seen from the partially plastic case of fig 1.9 that the strain in the outer fibres of the cross-section is $1/\alpha$ times the yield strain. For a mild steel having a plastic range as short as $10:1$, however, for which the strain in the outer fibres can reach ten times the yield strain before strain-hardening, the value of the moment of resistance can almost reach the fully plastic value before strain-hardening sets in. Setting $\alpha = 0\cdot1$ in equation (1.2) it will be seen that the factor in brackets differs from unity by only $\frac{1}{3}\%$. Thus although the theory requires the impossible condition that infinite strains be developed, it is sufficiently accurate to assume that a plastic hinge can form, and can undergo indefinite rotation under a constant value of resisting moment, M_p.

Equations (1.2) and (1.3) may be combined to give the general expression for the moment of resistance in the partially plastic state:

$$M_R = M_p(1 - \tfrac{1}{3}\alpha^2). \tag{1.4}$$

From this equation may be determined the shape of the plastic zones for some simple beams of rectangular cross-section. For example, the simply-supported beam of fig. 1.10 carries a central point load just

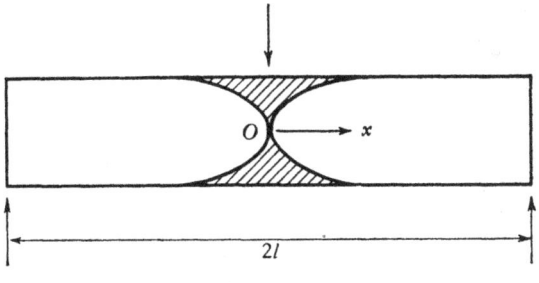

Fig. 1.10

sufficient in magnitude to cause the full plastic moment to be developed. If the span of the beam is $2l$, and the origin is taken at the centre, it will be seen that the bending moment at any section x of the beam $(x \geqslant 0)$ is given by

$$M = M_p\left(1 - \frac{x}{l}\right). \tag{1.5}$$

Equating the two expressions (1.4) and (1.5), $\alpha = \sqrt{(3x/l)}$, and the plastic zones will be as sketched in fig. 1.10. The plastic zones just meet under the central point load, and fall away sharply; the plastic hinge is confined to an infinitesimal length. In practice, rotation of this central hinge will cause the material to strain-harden, and the bending moment can increase above the nominal fully plastic value. This in turn will lead to a spread of the plastic zones from the hinge position, so that the hinge effectively extends over a short but finite distance.

The behaviour of the same beam subjected to a uniformly distributed load, fig. 1.11, is similar, but the plastic zones are larger. At collapse, the bending moments in the beam may be written

$$M = M_p \left(1 - \frac{x^2}{l^2} \right),$$
(1.6)

and, equating this expression to (1.4), it will be seen that $\alpha = \sqrt{3}(x/l)$. This beam will not strain-harden so quickly as that of fig. 1.10, and the slope of the strain-hardening characteristic, BC in fig. 1.2, will be correspondingly less.

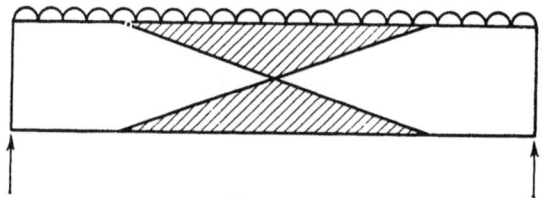

Fig. 1.11

Corresponding to the general expression for moment of resistance of a beam in the partially plastic state, equation (1.4), the curvature of the cross-section can also be computed in terms of the parameter α (fig. 1.9). The usual elastic formulae must hold for the central elastic core of depth $2\alpha d$, so that the curvature κ may be written

$$\kappa = \frac{1}{\alpha} \frac{\sigma_0}{Ed},$$
(1.7)

where E is the elastic modulus of the material. Equations (1.4) and (1.7) together are parametric expressions for the moment–curvature relationship in the elastic–plastic range, α having any value between 1 and 0. The slope of the curve is given by

$$\frac{\mathrm{d}M_R}{\mathrm{d}\kappa} = EI\alpha^3,$$
(1.8)

so that, for $\alpha = 1$, the elastic–plastic portion of the curve joins smoothly with the elastic line. Equations (1.4) and (1.7) are plotted in fig. 1.12, and the moment–curvature relationship of fig. 1.12 is close to that observed, fig. 1.5.

A similar analysis to that given above for the rectangular beam may be made for the I-section. A Universal Beam has 3° tapered flanges and

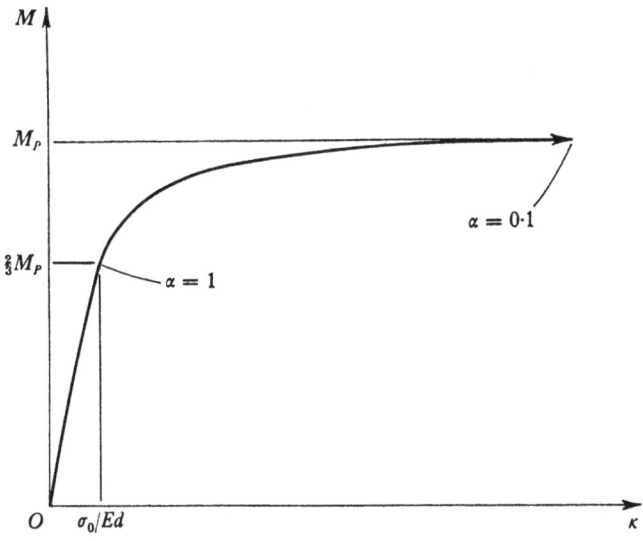

Fig. 1.12

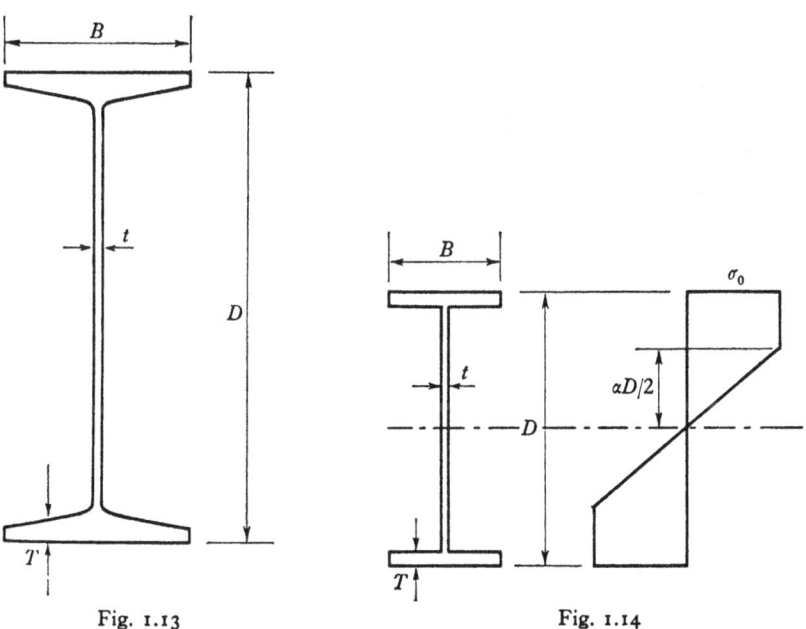

Fig. 1.13   Fig. 1.14

I. THE PLASTIC HINGE

a typical profile is shown in fig. 1.13. The tapered flanges will be replaced, for the purpose of analysis, by rectangular flanges of uniform thickness T, fig. 1.14, and it may be shown that the moment of resistance M_R in the partially plastic state is given by

$$M_R = \left[BT(D-T) + \left(\frac{D}{2}-T\right)^2 t - \tfrac{1}{12}D^2 t\alpha^2 \right] \sigma_0. \qquad (1.9)$$

In this expression, the value of α lies between $(D-2T)/D$ and zero; a separate analysis must be made for the case in which yielding of the cross-section has only just started, so that some of each flange is still elastic.

In equation (1.9), the contributions of the flanges and of the web are clearly apparent. As a numerical example, values will be taken from the section tables for a 12×5 Universal Beam at 48 kg/m;† for this section $B = 125$ mm, $D = 310$ mm, $T = 14.0$ mm, $t = 8.9$ mm. Thus equation (1.9) becomes

$$M_R = (518\,000 + 177\,000 - 71\,000\alpha^2)\,\sigma_0$$
$$= (695\,000 - 71\,000\alpha^2)\,\sigma_0, \qquad (1.10)$$

where the terms in brackets are in mm³.

The value of Z_p for this section may be determined more accurately by graphical means, and the value 705 000 mm³ is given in the section tables; an error of 1 % has been introduced by replacing the tapered flanges by rectangles of the same mean thickness. A closer match to the section properties may be obtained by allowing the dimensions marked in figs 1.13 and 1.14 to differ slightly, and formulae given in the section tables (for example, for computing the effects of axial load), are derived in this more accurate way.

The *elastic* modulus of the cross-section of fig. 1.14 is given to a close approximation by

$$Z_e = \frac{1}{D}\left[\{BT(D-T)^2 + \tfrac{1}{3}BT^3\} + \frac{4}{3}\left(\frac{D}{2}-T\right)^3 t \right] \qquad (1.11)$$

Numerically, for the 12×5 UB 48 kg, this expression gives

$$Z_e = 495\,000 + 107\,000 = 602\,000 \text{ mm}^3,$$

compared with the value 611 000 mm³ given in the section tables.

† At the time of printing, the *serial size* of a Universal Beam corresponds to the nominal dimensions in inches; thus the 12×5 UB considered has actual overall dimensions 12.22×4.93 in, equivalent to 310×125 mm.

The shape factor ν for the 12×5 UB 48 kg is given by $705/611 = 1\cdot15$, and this value is typical of those derived for all the Universal Beam sections. Thus, in a bend test of one of these sections, yield will not occur until the bending moment reaches the value $0\cdot87M_p$, and the moment–curvature characteristic for the cross-section will be as shown schematically in fig. 1.15, with the curvature increasing very rapidly after yield. (It may be noted that, as for the rectangular section, the curvature may be determined in terms of the parameter α; from fig. 1.14, $\kappa = 2\sigma_0/ED\alpha$.)

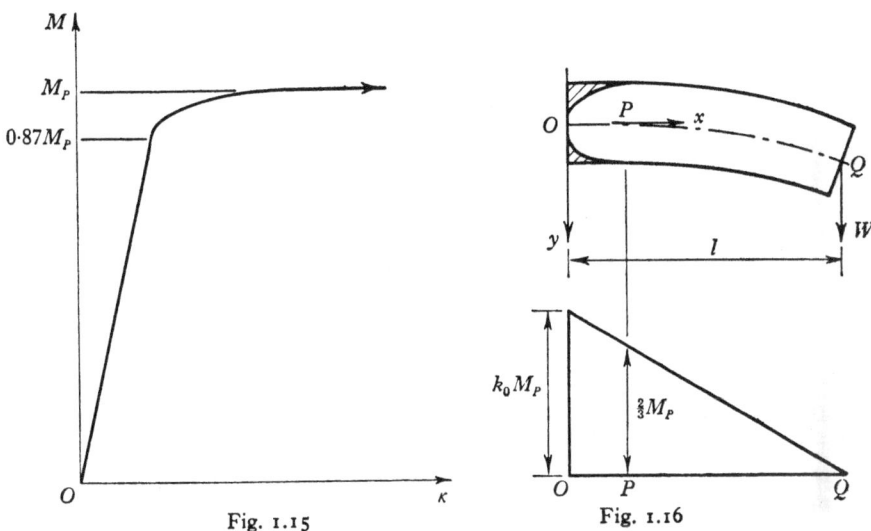

Fig. 1.15   Fig. 1.16

As a final example of the elastic–plastic behaviour of a steel beam, the load-deflexion curve will be calculated for a simply-supported beam, of rectangular cross-section, carrying a central point load. Considering first a cantilever of length l, fig. 1.16, it will be assumed that the tip load W is large enough so that a portion OP of the cantilever is partially plastic. Denoting the bending moment at the root of the cantilever by $k_0 M_p$, where

$$k_0 = Wl/M_p, \qquad (1.12)$$

it will be seen that the bending moment M_x at any section x will be given by

$$M_x = k_0\left(1 - \frac{x}{l}\right)M_p. \qquad (1.13)$$

Now from equation (1.4),

$$M_x = (1 - \tfrac{1}{3}\alpha^2)\,M_p, \qquad (1.14)$$

13

so that, equating these last two expressions, a relationship may be found between α and x which holds for the plastic region OP of the beam.

Equation (1.7) for the same region may be written

$$\frac{d^2y}{dx^2} = \frac{1}{\alpha}\frac{\sigma_0}{Ed}, \tag{1.15}$$

and this is the differential equation of bending for the plastic portion of the cantilever, for which α lies in the range $1 \geqslant \alpha \geqslant \alpha_0$, the value of α_0 being given from equations (1.13) and (1.14) by

$$1 - \tfrac{1}{3}\alpha_0^2 = k_0. \tag{1.16}$$

Equation (1.15) may be integrated to give the slope at any cross-section within the length OP:

$$\frac{dy}{dx} = \int_0^x \frac{1}{\alpha}\frac{\sigma_0}{Ed}\,dx = \int_{\alpha_0}^\alpha \frac{2}{3}\frac{\sigma_0}{Ed}\frac{l}{k_0}\,d\alpha = \frac{2}{3}\frac{\sigma_0}{Ed}\frac{l}{k_0}(\alpha - \alpha_0). \tag{1.17}$$

A second integration gives

$$y = \int_0^x \frac{2}{3}\frac{\sigma_0}{Ed}\frac{l}{k_0}(\alpha - \alpha_0)\,dx = \int_{\alpha_0}^\alpha \frac{4}{9}\frac{\sigma_0}{Ed}\frac{l^2}{k_0^2}(\alpha^2 - \alpha_0\alpha)\,d\alpha$$

$$= \frac{2}{27}\frac{\sigma_0}{Ed}\frac{l^2}{k_0^2}(2\alpha^3 - 3\alpha_0\alpha^2 + \alpha_0^3). \tag{1.18}$$

The slope and deflexion at P, where $M_x = \tfrac{2}{3}M_p$, can be found by setting $\alpha = 1$ in equations (1.17) and (1.18); the usual elastic solution for the portion PQ of the cantilever can then be added to give the deflexion at the tip Q.

These results may be used immediately to compute the central deflexion of the simply-supported beam of fig. 1.17:

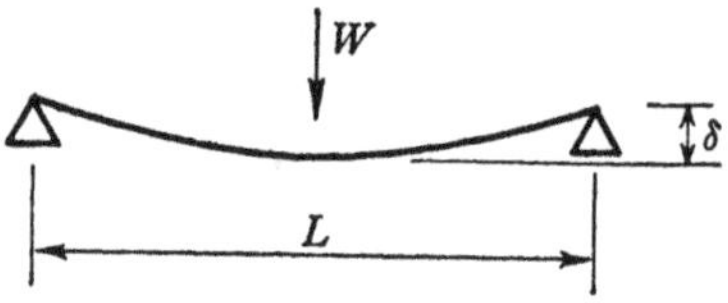

$$\delta = \frac{1}{54}\frac{L^2}{k_0^2}\frac{\sigma_0}{Ed}(10 - 9\alpha_0 + \alpha_0^3), \tag{1.19}$$

Fig. 1.17

where

$$k_0 = 1 - \tfrac{1}{3}\alpha_0^2 = WL/4M_p. \tag{1.20}$$

Equation (1.19) may be rearranged to give, in terms of the usual elastic flexural stiffness EI,

$$\delta = \frac{WL^3}{48EI}\frac{1}{2(\tfrac{3}{2}k_0)^3}(10 - 9\alpha_0 + \alpha_0^3). \tag{1.21}$$

At first yield, $k_0 = \tfrac{2}{3}$ and $\alpha_0 = 1$; equation (1.21) then gives the correct limiting elastic deflexion $WL^3/48EI$, where $W = \tfrac{8}{3}M_p/L$ from equation (1.20).

At collapse ($k_0 = 1$ and $\alpha_0 = 0$) the collapse load $W_c = 4M_p/L$ is reached at the finite deflexion

$$\delta_c = \frac{40}{27}\frac{W_c L^3}{48EI};$$
(1.22)

on substituting for the value of W_c, this deflexion is seen to be 20/9 times as large as the deflexion at first yield. Equation (1.22) is shown as portion AB of the curve in fig. 1.18; as mentioned above, increased deflexions beyond the point B must be accompanied in practice by strain hardening and by a slight 'spread' of the central plastic hinge.

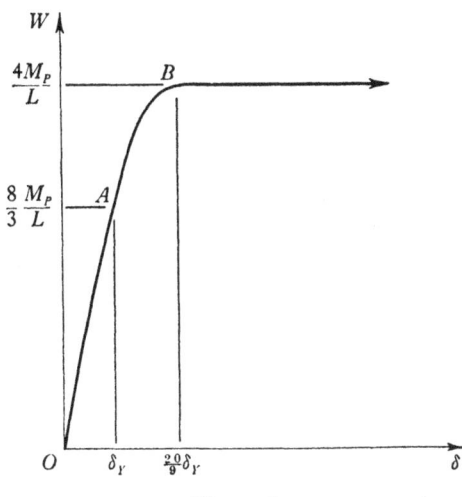

Fig. 1.18

This simple example demonstrates that it is possible to reproduce the essential features of a bend test using the simple theory of bending and material properties derived from a tensile test. The collapse load of the simply-supported beam of fig. 1.17 is given by the expression $WL = 4M_p$. Assuming that deflexions are, in the first instance, not of prime importance, then the collapse analysis requires a knowledge only of the value of the full plastic moment; knowing the value of the yield stress of the material, this in turn implies a knowledge only of the plastic modulus of the cross-section.

1.4 Plastic moduli of sections having at least one axis of symmetry

It was tacitly assumed in the last diagram of fig. 1.9 that the zero stress axis coincided with the usual neutral axis of the elastic theory of bending.

This assumption is valid only if the cross-section has two axes of symmetry. A general cross-section having only one axis of symmetry is shown in fig. 1.19; the section is being bent about an axis at right angles to this axis of symmetry. The fully plastic distribution of stress will be as shown, with the portion of the cross-section lying above the zero stress axis being subjected to the uniform yield stress, which will be compressive for a sagging bending moment, while the portion below the zero stress axis is yielding uniformly in tension. The total force on the

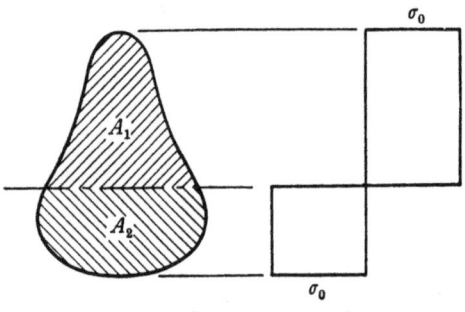

Fig. 1.19

top half of the beam is thus $\sigma_0 A_1$, and on the bottom half $\sigma_0 A_2$; in the absence of axial load, these two quantities must be equal, so that $A_1 = A_2$. The zero-stress axis is, in fact, an *equal area* axis; in general, this axis will not pass through the centre of gravity of the whole cross-section, as does the elastic neutral axis. By writing the general equation for elastic–plastic bending, it is possible to trace the gradual shift of the zero-stress axis as the bending moment is increased, from the centroidal position of elastic theory to the equal-area position of full plasticity.

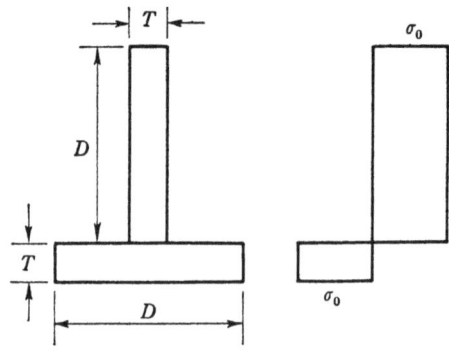

Fig. 1.20

16

The T-section composed of two equal rectangular plates shown in fig. 1.20 will have the fully plastic stress distribution shown. The force on each rectangle has value $\sigma_0 DT$, and the lever arm is $\frac{1}{2}(D+T)$, so that the plastic modulus is given by

$$Z_p = \tfrac{1}{2}DT(D+T). \tag{1.23}$$

This value may be compared with the elastic modulus

$$Z_e = \left(\frac{DT}{3D+T}\right)[\tfrac{1}{3}(D^2+T^2)+\tfrac{1}{2}(D+T)^2]. \tag{1.24}$$

Thus for $D/T = 8$, say, the shape factor may be determined as

$$\nu = Z_p/Z_e = \tfrac{675}{373} = 1\cdot81.$$

1.5 Design of cross-sections; cover plates

The calculation of plastic moduli is essentially a matter of computing the first moment of area of the cross-section about the zero-stress axis. There is thus no formal difficulty, although the calculations may be tedious. Occasionally a doubly symmetric standard rolled section may be modified, either by the addition of cover plates or by the curtailment of flanges, into a section having only one axis of symmetry; in these cases, it is possible to shorten the calculations by casting them in a standard form.

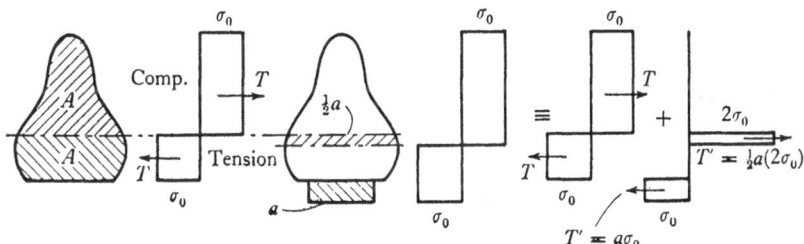

Fig. 1.21

Suppose a section having at least one axis of symmetry, and of area $2A$, is strengthened by the addition of a single cover plate of area a, fig. 1.21. Since the zero stress axis must be the equal area axis for the strengthened cross-section, it is clear from the diagram that the axis must shift from its original position so that an area $a/2$ of the original cross-section moves from the tension side to the compression side of the axis. The total force on the cover plate in the fully plastic state is $a\sigma_0$; from fig. 1.21 it will be seen that the final fully plastic stress distribution may be thought of as consisting of two parts. The original full plastic moment is increased by a moment due to the flange plate force $a\sigma_0$ and a force due to the fictitious stress $2\sigma_0$ acting on the area $a/2$ that has been transferred from tension to compression.

As a first example, suppose the 12×5 UB 48 kg is strengthened by the addition of a small cover plate to the lower flange, the dimensions being 150×12 mm, fig. 1.22. The area a is 1800 mm², and since the

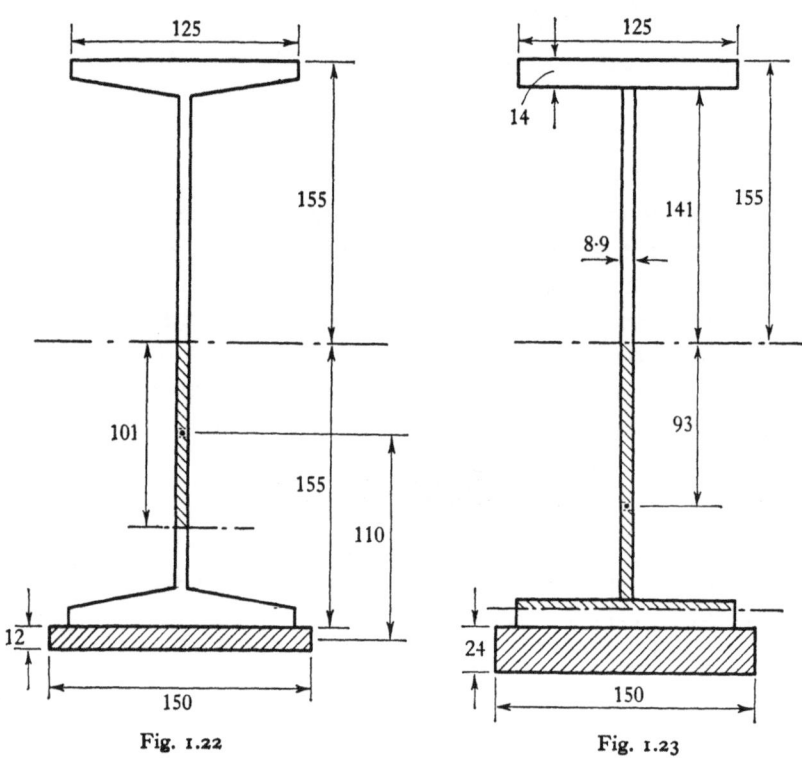

Fig. 1.22   Fig. 1.23

web thickness is 8·9 mm, the zero-stress axis must shift through a distance $900/8\cdot9 = 101$ mm. The lever arm for computing the additional plastic modulus is therefore $(155 + 6 - 51) = 110$ mm, and the additional modulus is $(1800)(110) = 198\,000$ mm³. This value must be added to the value 705 000 from the section tables to give a total 903 000 mm³.

This value is exact; the addition of a larger cover plate, say 150×24 mm, fig. 1.23, will shift the zero stress axis into the flange, and a graphical method allowing for fillets and flange taper must be used for an extremely accurate answer. Alternatively, the approximation of three rectangles, using dimensions from the section tables, will be found to give sufficient accuracy for the purpose of design. The area of the flange plate is $150 \times 24 = 3600$ mm², so that an area of 1800 mm² must shift from tension to compression. The effective area of half the web is $(141)(8\cdot9) = 1253$ mm²; thus a depth $(1800 - 1253)/(125)$ $= 4\cdot38$ mm of the flange must also lie above the zero-stress axis. The centre of area of the 'transferred' portion of the cross-section lies 93 mm from the original axis of bending, so that the lever arm for the

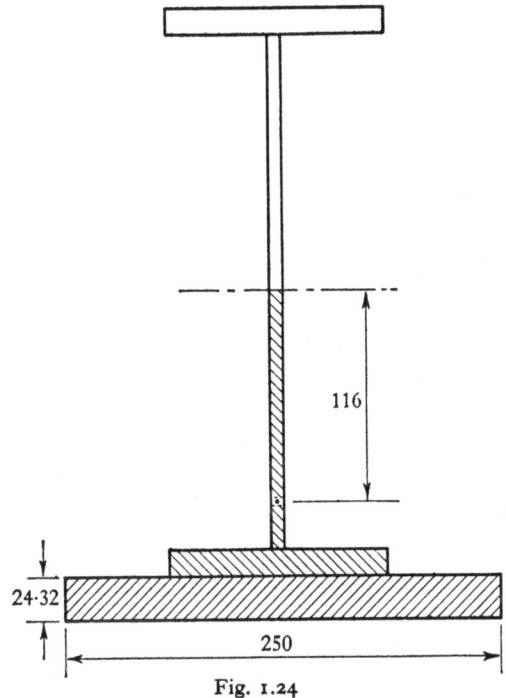

Fig. 1.24

increase in plastic modulus is $(155 + 12 - 93) = 74$ mm. The increase in modulus is thus $(3600)(74) = 266000$ mm³, giving a total

$$Z_p = 705\,000 + 266\,000 = 971\,000 \text{ mm}^3.$$

As a final, somewhat artificial, example, fig. 1.24, a $250 \times 24\cdot32$ mm cover plate has an area exactly equal to that of the original I-section. Repeating the calculations for this new cover plate, the centre of area of half the original section lies 116 mm from the original centre line, so that the lever arm for the additional modulus is $(155 + 12 - 116) = 51$ mm, the increase in modulus being $(6080)(51) = 310\,000$ mm^3. The total modulus is thus $705\,000 + 310\,000 = 1\,015\,000$ mm^3, which may be checked directly by a simple calculation. Since the whole of the I-section is in compression, and the whole of the cover plate in tension, the plastic modulus of the strengthened section is

$$(6080)(155 + 12) = 1\,015\,000 \text{ mm}^3,$$

as before.

1.6 Effect of axial load

If an axial load is applied to the cross-section of a short column, then the load will give rise to a uniform compressive stress over the section. Addition of a small bending moment will then produce a linear variation of elastic stress across the section, fig. 1.25. Further increase of bending

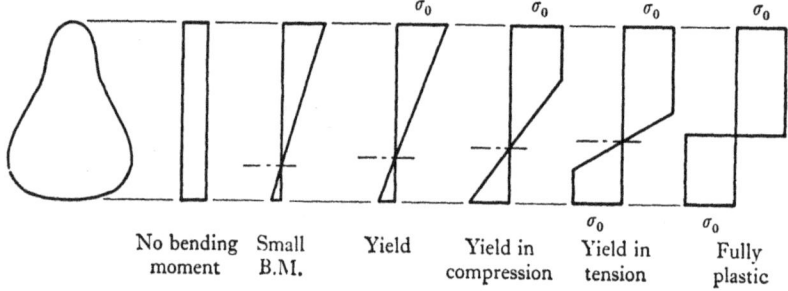

No bending moment   Small B.M.   Yield   Yield in compression   Yield in tension   Fully plastic

Fig. 1.25

moment, with the axial load remaining constant, will eventually cause yield to occur on one face of the section, followed by yield on the other face, and eventual yield of the whole cross-section. During this process the zero-stress axis, which lies outside the section for very small values of bending moment, shifts progressively towards the final position in the fully plastic state.

It is this full plastic state that is of interest, and to illustrate the effect of axial load on the value of the full plastic moment, a rectangular cross-section will first be examined, fig. 1.26. An axial load P is supposed to act at the centre line of the cross-section, in addition to a bending moment M_p sufficient to cause the whole section to be fully plastic. In order that there shall be a net resultant force P across the cross-section, the zero-stress axis must shift from the centre line, by an amount βd, say.

It will be seen that the fully plastic stress distribution may be regarded as being composed of two parts; that is, on the original fully plastic distribution in the absence of axial load may be superimposed a fictitious distribution involving stresses of magnitude $2\sigma_0$. This second distribution must be equivalent to a total force P, and it will reduce the value of the full plastic moment by an amount $(P)(\tfrac{1}{2}\beta d)$. Thus from fig. 1.26,

$$P = 2\beta bd\sigma_0 = \beta P_0, \\ M_p = M_{p_0} - P(\tfrac{1}{2}\beta d) = (1 - \beta^2)\,M_{p_0}, \qquad (1.25)$$

where P_0 is the 'squash load' of the cross-section in the absence of bending moment, and M_{p_0} is the value of the full plastic moment $(bd^2\sigma_0)$ in the absence of axial load.

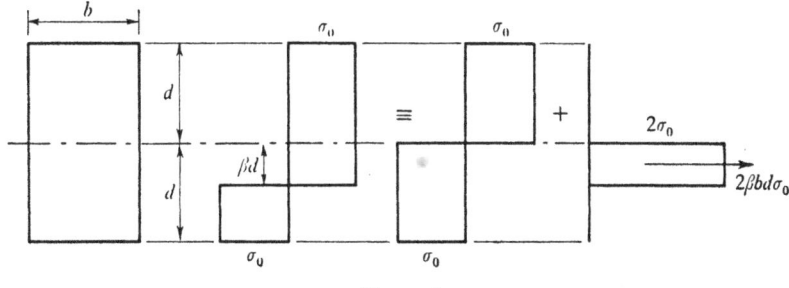

Fig. 1.26

Equations (1.25) may be combined to give

$$\left(\frac{M_p}{M_{p_0}}\right) + \left(\frac{P}{P_0}\right)^2 = 1 \qquad (1.26)$$

which gives the characteristic curve of fig. 1.27. This curve is doubly symmetric; the load P has been thought of as being compressive, but exactly the same analysis holds for tensile loads, and for bending in the reversed direction; thus the complete plot will be as shown. For small values of axial load, there is very little reduction in the value of M_p.

Figure 1.27 is an example of an important concept in plastic theory, that of the *yield surface*. A point in the plane of the figure represents, for a given rectangular section having known values of M_{p_0} and P_0, a certain combination of bending moment and axial load. If the point lies *within* the convex boundary, then the combination is one that can be carried by the cross-section. A point *on* the boundary of the convex yield surface represents a combination of bending moment and axial load that just causes the section to become fully plastic. A point *outside* the convex region represents an impossible state.

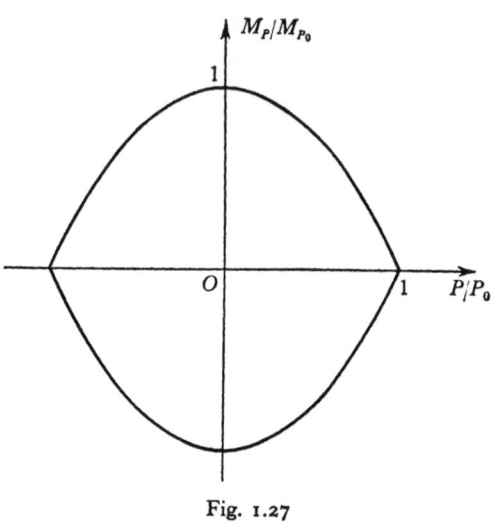

Fig. 1.27

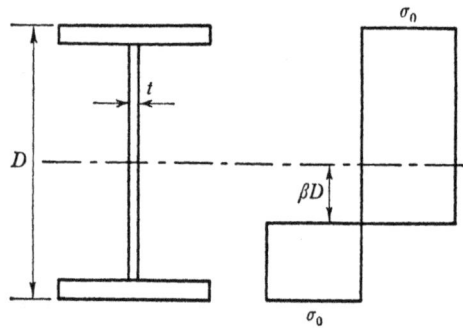

Fig. 1.28

The analysis given above for the rectangular section can be applied with little change to determine the effect of a small axial load on a rolled I-section. From fig. 1.28 it will be seen that

$$M_p = M_{p_0} - \beta^2 t D^2 \sigma_0,$$
$$P = 2\beta t D \sigma_0. \qquad (1.27)$$

If the mean axial stress is denoted by p, so that $P = pA$ where A is the total cross-sectional area, and if $n = p/\sigma_0$, then equations (1.27) may be combined to give

$$Z_p = Z_{p_0} - \beta^2 t D^2$$

$$= Z_{p_0} - \left(\frac{A^2}{4t}\right) n^2. \tag{1.28}$$

For the 12×5 UB 48 kg, $A = 6080$ mm² and $t = 8.9$ mm, so that equation (1.28) becomes

$$Z_p = 705\,000 - 1\,038\,000n^2. \tag{1.29}$$

Expressions similar to (1.29) are given in the standard section tables, and hold only for relatively small values of axial load. At larger values of n the zero-stress axis moves from the web into one flange, and a new analysis must be made, using either graphical methods or the idealized section composed of three rectangles. For the 12×5 UB 48 kg, the following expression may be derived for the value Z_p when $n > 0.412$:

$$Z_p = 74\,480\,(1 - n)(11.66 + n). \tag{1.30}$$

Equations (1.29) and (1.30) may be plotted to give the curve of fig. 1.29, which is one quadrant of a complete yield surface of the type shown in fig. 1.27.

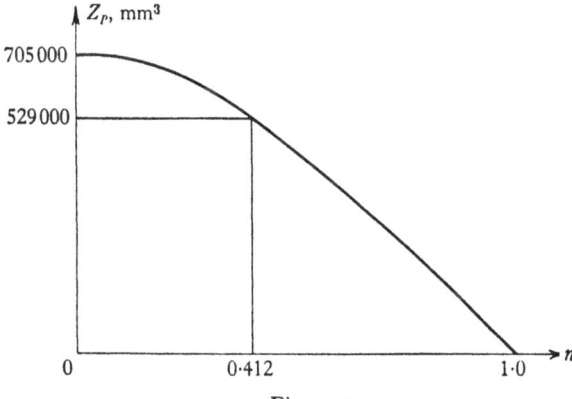

Fig. 1.29

For the more general case of a cross-section having at least one axis of symmetry, it may be supposed that the axial load P (acting in the plane of the original zero-stress axis) causes the zero-stress axis to shift so that an area a is transferred from tension to compression (or vice versa). From fig. 1.30, if the centre of the transferred area a is distant $\bar{y}$ from the original zero-stress axis,

$$P = 2\sigma_0 a, \\ M_p = M_{p_0} - P\bar{y},$$

(1.31)

that is,

$$Z_p = Z_{p_0} - 2a\bar{y},$$

(1.32)

which is a simple general expression for computing the reduction in plastic modulus due to axial load.

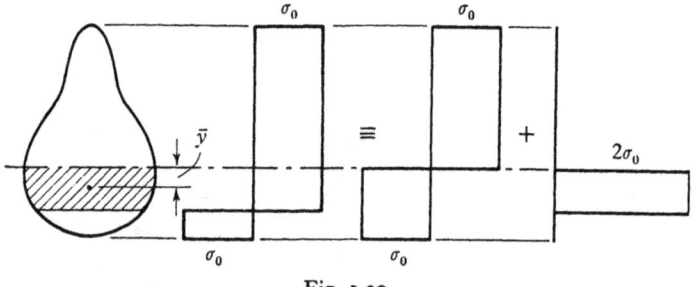

Fig. 1.30

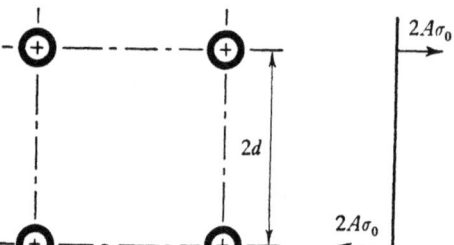

Fig. 1.31

As a further example of the effect of axial load, suppose that a column is made up of four steel tubes (fig. 1.31) of negligible linear dimensions but each of finite cross-sectional area A; the tubes are adequately braced together to form a beam of square cross-section. In the absence of axial load, the full plastic moment of the beam, neglecting any contribution from the bracing, is

$$M_{p_0} = 4Ad\sigma_0.$$

(1.33)

If now the cross-section is subjected to a compressive load P acting in the plane of two of the tubes, fig. 1.32, it will be seen that the net forces in those two tubes must fall below the yield value $2A\sigma_0$. The zero-stress axis must intersect these tubes so that part of the cross-section is yielding in compression, while the remainder yields in tension. The value of the resisting moment remains unchanged at

$$M_p = M_{p_0} = 4Ad\sigma_0,$$

the value of equation (1.33). The value of the compressive load P cannot, of course, exceed $4A\sigma_0$, corresponding to all four tubes yielding in compression.

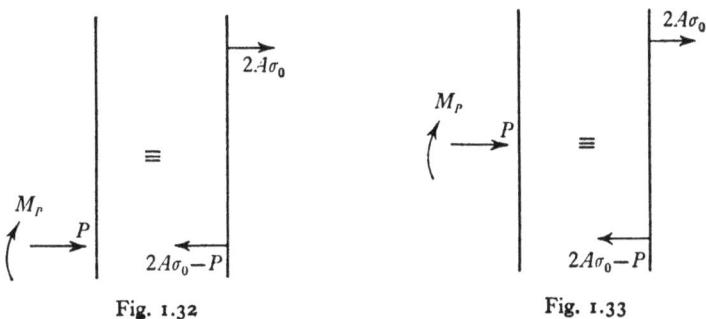

Fig. 1.32   Fig. 1.33

If the compressive load P is specified more conventionally with respect to the zero-stress axis (axis of symmetry in this case), it will be seen from fig. 1.33 that

$$M_p = M_{p_0} - Pd. \qquad (1.34)$$

It is therefore important, when computing reductions in plastic modulus due to axial load, to specify the line of action of that load. If the line of action is taken to be the original zero-stress axis, then equations (1.31) show that the full plastic moment is always reduced by the axial load. If the load is not specified with respect to that axis, then, effectively, the axial load is combined with an additional bending moment. If the sign of the additional bending moment is favourable, then the full plastic moment may apparently not be reduced, as above, and may, indeed, appear to increase. This increase is, however, purely illusory.

The effect may perhaps be made clear by returning to the example of the T-section composed of two equal rectangles, fig. 1.20. In fig. 1.34 are shown the two cases of the zero-stress axis lying in one or other of the rectangles. If the sense of the applied bending moment is such that the top of the T-section is in compression, then fig. 1.34(*b*) corresponds to a compressive load

$$P = -2\alpha TD\sigma_0 = -\alpha P_0,$$

(1.35)

where P_0 is the numerical value of the squash load, and the negative sign indicates compression. Similarly, the axial load corresponding to fig. 1.34(*c*) is tensile, of value

$$P = 2\beta TD\sigma_0 = \beta P_0.$$

(1.36)

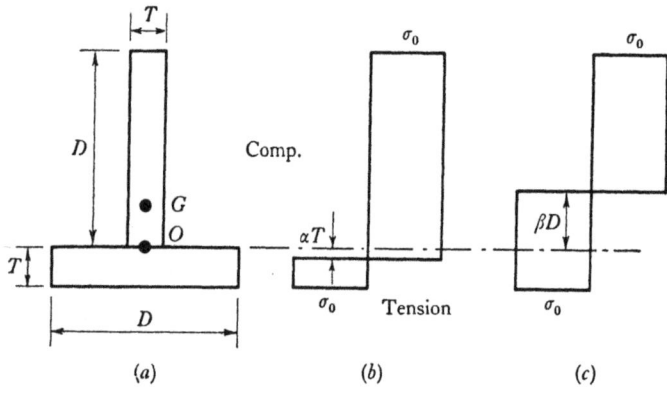

Fig. 1.34

If, for simplicity, the ratio T/D is taken as small, and the load P acts at the position of the original zero-stress axis (O in fig. 1.34(*a*)), then, corresponding to equations (1.35) and (1.36),

for
$$\frac{P}{P_0} \leqslant 0, \quad \frac{M_p}{M_{p_0}} = 1,$$

and for
$$\frac{P}{P_0} \geqslant 0, \quad \left(\frac{M_p}{M_{p_0}}\right) + 2\left(\frac{P}{P_0}\right)^2 = 1.$$

(1.37)

These equations are plotted in fig. 1.35; by considering bending to occur in the opposite sense, the skew symmetric figure may be completed. If now the load P acts through the centre of gravity G of the cross-section, instead of through O, fig. 1.34(*a*), the equations for the value of the full plastic moment are,

for
$$\frac{P}{P_0} \leqslant 0, \quad \frac{M_p}{M_{p_0}} - \frac{P}{P_0} = 1,$$

and for
$$\frac{P}{P_0} \geqslant 0, \quad \frac{M_p}{M_{p_0}} + \frac{P}{P_0}\left(2\frac{P}{P_0} - 1\right) = 1.$$

(1.38)

26

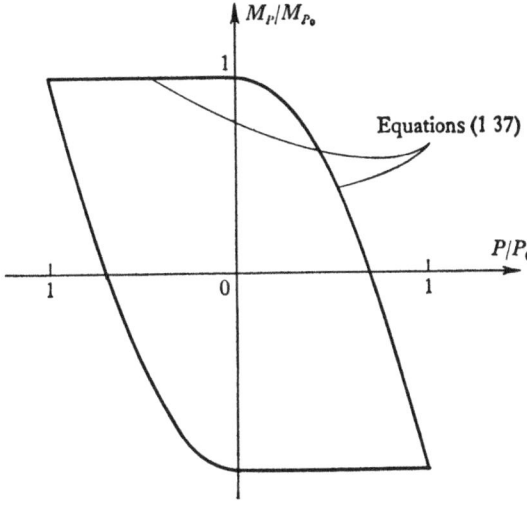

Fig. 1.35

These equations are plotted in fig. 1.36. Comparing with fig. 1.35, it will be seen that one diagram can be derived from the other by a simple 'shear', and this is a general property. In this case, equations (1.38) may be derived from equations (1.37) simply by subtracting P/P_0 from the left-hand sides.

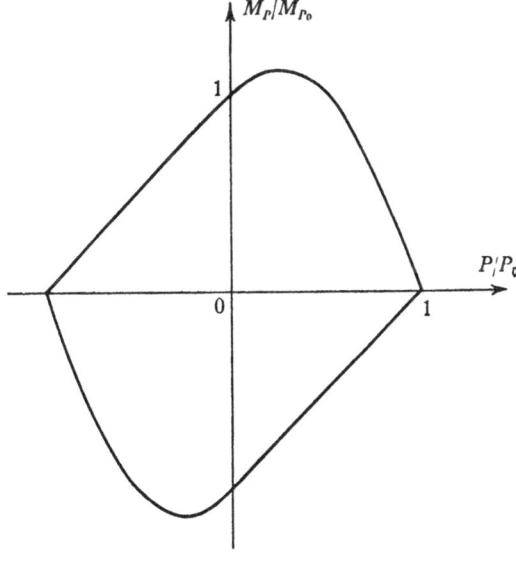

Fig. 1.36

Diagrams of the form of fig. 1.36 are perhaps more natural; the squash load must act through the centre of gravity G, and when the full squash load acts, there can be no bending moment about an axis through G (M_p/M_{p_0} is zero in fig. 1.36 for $P/P_0 = \pm 1$). However, using the axis through G for reference has the anomaly that a small axial load can apparently lead to an increase in bending moment. In diagrams like fig. 1.35, however, in which the axial load is referred to the original zero-stress axis, the axial load cannot lead to an increase in the value of full plastic moment.

For a section bent about an axis of symmetry the zero-stress axis passes through the centre of gravity, and these difficulties do not arise.

1.7 Effect of shear force

The effect of shear force on a beam of general cross-section is much more complex than that of axial load. With the latter, the resulting stresses could be superimposed directly on the bending stresses, since they were all longitudinal. Shear combined with bending gives rise, however, to a two-dimensional stress system, and a general discussion of plastic behaviour under these conditions is outside the scope of this volume. The special case of the I-section may be dealt with, however, by an approximate method. It is commonly assumed in elastic design that the shear stresses are uniformly distributed over the web of an I-section, the flanges not contributing at all to the carrying of the shear force. If these same assumptions are made in the plastic analysis of the problem, then an empirical solution may be obtained which gives good agreement with experimental results.

It was seen above, equation (1.9), that the full plastic moment of an I-section, fig. 1.37, is given by

$$M_p = BT(D-T)\sigma_0 + \left(\frac{D}{2} - T\right)^2 t\sigma_0$$

$$= M_f + M_{w_0} \quad \text{say,} \tag{1.39}$$

where M_f and M_{w_0} are the respective contributions from the flanges and from the web. Suppose now that a shearing force F acts on the web causing a uniform shear stress τ, so that

$$F = (D-2T)t\tau. \tag{1.40}$$

If the web remains fully plastic, then the longitudinal stress σ in the web available for resisting the bending moment will be reduced below the

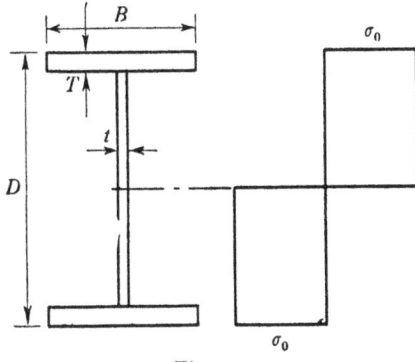

Fig. 1.37

value σ_0. There are several criteria for yield under combined stress systems; the criterion which fits best the experimental results for this problem is due to Mises, and in this simple case gives

$$\sigma^2 + 3\tau^2 = \sigma_0^2. \tag{1.41}$$

The stress distribution over the cross-section under combined shear and bending sufficient to cause full plasticity will be as shown in fig. 1.38. The contribution from the flanges remains unaltered, but the contribution from the web is reduced in the ratio σ/σ_0.

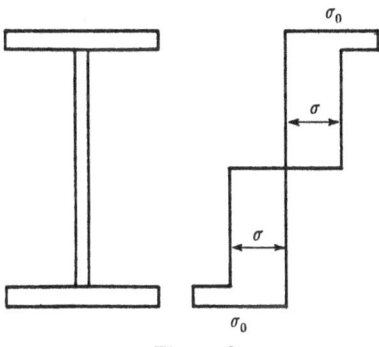

Fig. 1.38

This analysis may be applied to a built-up plate girder or, with the usual approximation of three rectangles, to a standard rolled section. As an example, suppose the 12 × 5 UB 48 kg is subjected to a shearing force of 300 kN, and has a yield stress $\sigma_0 = 250$ N/mm². From equation (1.40), using the dimensions given in the section tables

$$\tau = (300\,000)/(282)(8 \cdot 9) = 119 \cdot 4 \text{ N/mm}^2,$$

so that, from equation (1.41), $\sigma/\sigma_0 = \sqrt{\{1 - 3(119\cdot4/250)^2\}} = 0\cdot562$. Equation (1.39) may be re-written

$$Z_{p_0} = Z_f + (141)^2(8\cdot9) = Z_f + 177\,000, \qquad (1.42)$$

so that, under the action of the shear force,

$$Z_p = Z_f + 177\,000\frac{\sigma}{\sigma_0} = Z_{p_0} - 177\,000\left(1 - \frac{\sigma}{\sigma_0}\right). \qquad (1.43)$$

Inserting the value $\sigma/\sigma_0 = 0\cdot562$, it will be seen that the original plastic modulus, $Z_{p_0} = 705\,000$ mm³, is reduced by $(177\,000)(0\cdot438) = 78\,000$ to the value $627\,000$ mm³.

The general relation between plastic modulus and shear force for this 12×5 UB 48 kg section is shown in fig. 1.39. For small values of shear force there is very little reduction in Z_p. When the whole of the web is yielding in shear at an average stress of $250/\sqrt{3} = 144$ N/mm², the flanges remain to contribute a plastic modulus $Z_f = 528\,000$ mm³.

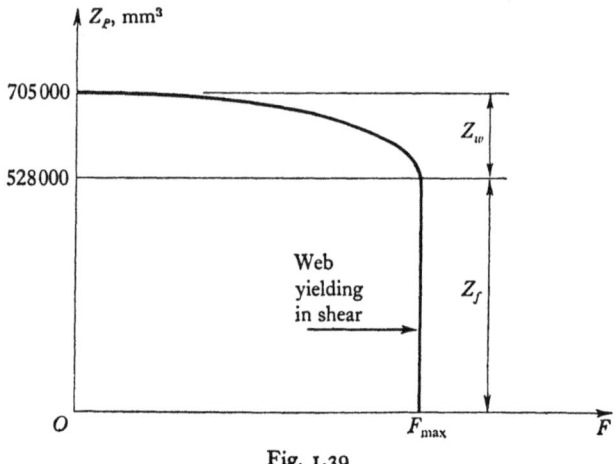

Fig. 1.39

Both axial load and shear force tend to lead to instability of the flanges and webs of I-section girders, if the depth to thickness ratios of these are too large; certain limits must therefore be imposed if full plasticity is to develop without the danger of local buckling. The limits depend on the square root of the yield stress of the material; for Grade 43 steel for which $\sigma_0 = 250$ N/mm², B/T (for the flanges) must not exceed 18 and $(D - 2T)/t$ (for the webs) must not exceed 85. The corresponding values for Grade 50 steel ($\sigma_0 \simeq 350$ N/mm²) are 15 and 70.

These limits will be reduced if the *combined* effects of shear and axial load are considered, and due attention must be paid to this point if, in any practical design, both shear force and axial load are high. The direct effect of shear force and axial load acting together on an I-section may be found by considering fig. 1.40. In this figure, a bending stress

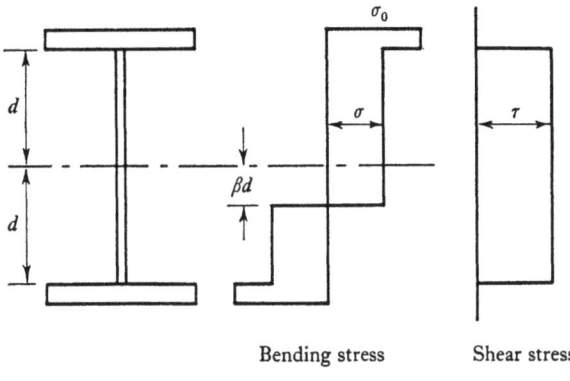

Bending stress   Shear stress

Fig. 1.40

distribution analogous to that of fig. 1.38 has been assumed; however, this distribution has been modified to allow for the effect of axial load (cf. fig. 1.28). The plastic section modulus of the web may be written directly by using equations (1.25) in conjunction with the analysis given above for the effect of shear force alone:

$$Z_w = Z_{w_0}(1 - \beta^2)\left(\frac{\sigma}{\sigma_0}\right), \qquad (1.44)$$

where
$$P = \beta P_0\left(\frac{\sigma}{\sigma_0}\right). \qquad (1.45)$$

In equation (1.45), P_0 is the squash load of the *web* alone. Using the yield condition, equation (1.41),

$$Z_w = \left[\sqrt{\{1 - 3(\tau/\sigma_0)^2\}} - \frac{(P/P_0)^2}{\sqrt{\{1 - 3(\tau/\sigma_0)^2\}}}\right] Z_{w_0}. \qquad (1.46)$$

In this equation, the mean web shearing stress τ is computed, as before, on the assumption that none of the shear force is carried by the flanges of the I-section.

More complex expressions result if the axial load P is so large that the zero stress axis moves out of the web into the flange of the I-section.

31

1.8 Bending about an inclined axis

So far, all the cross-sections whose plastic moduli have been determined have had at least one axis of symmetry, and have been bent about one of the two 'obvious' axes, corresponding to the principal axes of elastic theory. Suppose now that a symmetrical rectangular cross-section, $2a \times 2b$, is bent about an inclined axis, fig. 1.41. In elastic bending the neutral axis is not necessarily parallel to the axis of the applied bending moment. Similarly, in plastic bending the zero-stress axis for full plasticity will not necessarily be parallel to the axis of the bending moment.

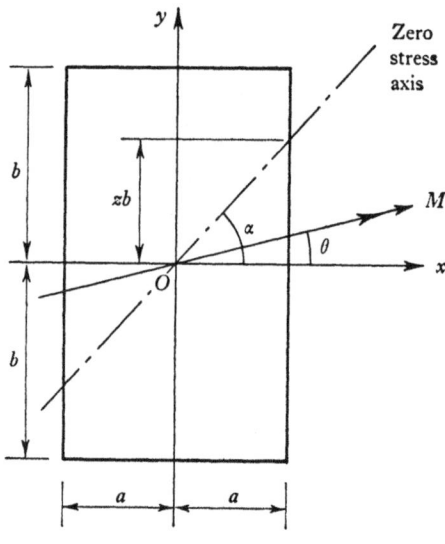

Fig. 1.41

In fig. 1.41 the axis of the applied bending moment M makes an angle θ with the direction of Ox, while the zero-stress axis makes an angle α with Ox, where, from the figure,

$$\tan\alpha = \frac{b}{a}z. \tag{1.47}$$

By taking moments about Ox and Oy in turn, it is found that the components M_x and M_y of the bending moment M are given by

$$\left. \begin{aligned} M_x &= M\cos\theta = 2ab^2\sigma_0(1 - \tfrac{1}{3}z^2), \\ M_y &= M\sin\theta = 2a^2b\sigma_0(\tfrac{2}{3}z). \end{aligned} \right\} \tag{1.48}$$

32

These expressions hold for $-1 \leqslant z \leqslant 1$, that is, for the value of $\tan\alpha$ less than b/a; similar expressions may be derived for the case when $\tan\alpha$ is greater than b/a.

Equations (1.48) are parametric expressions for the bending moments M_x and M_y; the equations can be rearranged to give

$$
\left.
\begin{aligned}
M^2 &= M_x^2 + M_y^2, \\
\tan\theta &= \frac{M_y}{M_x}.
\end{aligned}
\right\}
\qquad (1.49)
$$

A complete plot of equations (1.48), the similar expressions for $\tan\alpha$ greater than b/a, and of all these expressions with the signs reversed (bending in the opposite sense), is shown in fig. 1.42. This is the yield surface for the rectangular section acted upon by bending moments M_x and M_y. From equations (1.48) it will be seen that the vector OP from the origin to any point P on the curve gives the magnitude M of the total bending moment and the angle θ that its axis makes with the axis of M_x.

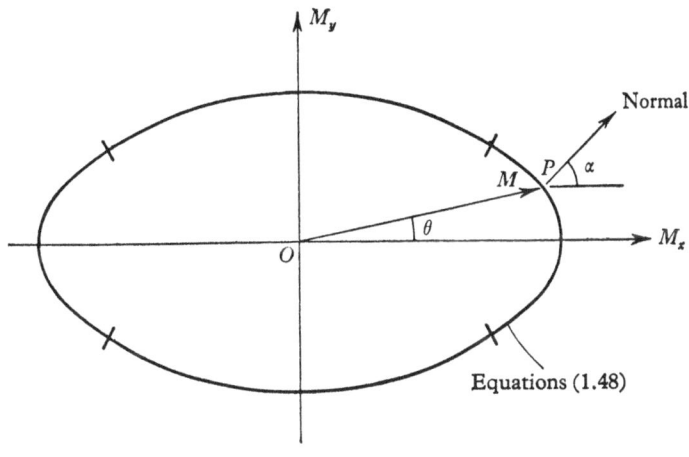

Fig. 1.42

A further general property of the yield surface concerns the direction of the normal to the surface at any point. In fig. 1.42, the normal at the point P is shown making an angle α with the direction of the axis of M_x. This is the same α as that of equation (1.47), as may be seen readily. The slope of the normal is $-\mathrm{d}M_x/\mathrm{d}M_y$, and, from equations (1.48), this is equal to bz/a.

33

This *normality* property is general. If deformation axes (e.g. hinge rotations or curvatures) are superimposed on the force axes, then the normal to the yield surface will indicate the relative proportions of the deformations.

From equation (1.49), the value of M is a maximum or minimum when

$$M_x\,dM_x + M_y\,dM_y = o,$$

or

$$\frac{M_y}{M_x} = -\frac{dM_x}{dM_y}, \tag{1.50}$$

that is, in general,

$$\tan\theta = \tan\alpha. \tag{1.51}$$

Thus the zero-stress axis and the axis of the applied bending moment will coincide when M is a maximum or minimum. In general, there will be two positions, which may be called the strong and weak principal axes of plastic bending, corresponding to the two elastic principal axes.

For a cross-section having two axes of symmetry, those axes are both the elastic principal axes and the strong and weak principal axes of plastic bending. A cross-section having one axis of symmetry only has already been discussed; for the T-section it was seen that the equal area axis did not pass through the centre of gravity, so that the elastic and plastic principal axes no longer coincided, although they were parallel.

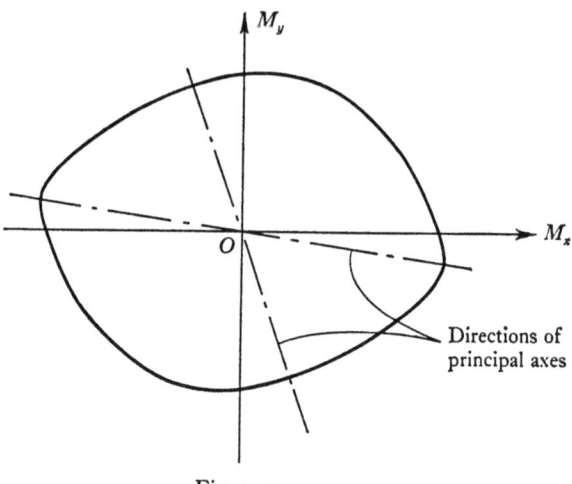

Fig. 1.43

The general unsymmetrical section will not be discussed in detail in this volume. However, it may be noted that the yield surface will, in general, be skew-symmetric, of the type sketched in fig. 1.43. The plastic principal axes will be located by the condition that the radius from the origin is normal to the yield surface. Thus the two plastic principal axes need not coincide at all with the corresponding elastic axes; indeed, they need not be at right angles. As an example, to be treated in volume 2, the plastic principal axes of an unequal angle section, ratio of leg length 4 to 3, intersect at about 80° rather than at right angles.

Care must be exercised, therefore, in any practical design which uses a section bent about an axis which is not one of symmetry.

1.9   The load factor

A structure is always designed with some margin of safety; that is, the working loads are less than the collapse loads. Suppose a simply-supported beam, fig. 1.44, is subjected to a set of working loads which can be specified in terms of one of their number, W. The bending-moment diagram can be drawn and the maximum bending moment M determined. If now all the loads are increased by a common factor λ, the maximum bending moment in the beam will increase by the same factor λ. Collapse will occur when the maximum bending moment $\lambda_c M$ reaches the full plastic value, due account being taken of any effects of shear force and axial load. The value λ_c is called the *collapse load factor*, that is, it is the ratio of the collapse load to the working load on the structure.

The primary function of a load factor is to ensure that a structure will be safe under service conditions. However, the load factor performs several duties, since margins of safety are required to cover uncertainties in the values of the loads, imperfections in workmanship, errors in design and fabrication, and so on. With so many variables, it is difficult to fix a value of the load factor on a theoretical basis, although a probability analysis can be of help.

However, recourse may be had to the experience gained with existing steel structures. Steel frames designed in accordance with the conventional elastic rules of BS 449† have shown themselves to be safe, and

---

† *British Standard 449: The Use of Structural Steel in Building.* The Specification is concerned in the main with 'simple design', that is, with the application of conventional elastic methods in a way which is sometimes no more than rule-of-thumb. However, 'fully rigid design' by plastic methods is permitted by the Standard, although no load factor is specified and no guidance is given. At the time of writing the following countries have equivalent Specifications which permit plastic design: Australia, Belgium, Canada, Czechoslovakia, Germany, Hungary, India, Japan, Norway, Portugal, Sweden, Switzerland, U.S.A.

plastic designs incorporating the minimum load factor implied by these rules should therefore also be safe. As will become apparent from the discussions of beams in chapter 2, this minimum load factor is that for a statically determinate member, such as a simply-supported beam. Its value can be found by considering the behaviour of the beam of fig. 1.44.

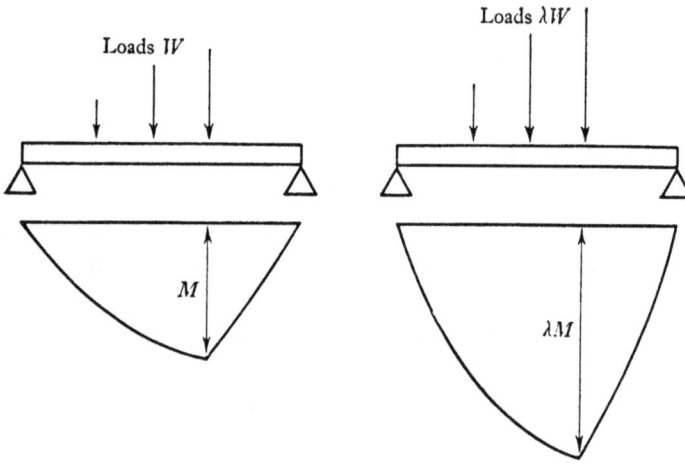

Fig. 1.44

An elastic design of the beam would be governed by the stress corresponding to the largest bending moment M; if the maximum permitted stress were σ_e, then the elastic section modulus would be given by

$$M = \sigma_e Z_e. \tag{1.52}$$

The collapse condition of this same beam is given by

$$\lambda_c M = M_p = \sigma_0 Z_p, \tag{1.53}$$

so that

$$\lambda_c = \frac{\sigma_0}{\sigma_e} \frac{Z_p}{Z_e} = \frac{\sigma_0}{\sigma_e} \nu. \tag{1.54}$$

If the design were in Grade 43 steel, then the ratio of the guaranteed yield stress to the maximum permitted bending stress would be about 1·52. Assuming that the shape factor ν for a rolled I-section is 1·15, then equation (1.54) gives

$$\lambda_c = (1\cdot52)(1\cdot15) = 1\cdot75. \tag{1.55}$$

BS 449 permits an increase of 25 % in stresses if such an increase is solely due to the effect of wind. This implies a reduction in load factor to 1·40, and represents the statistical unlikelihood of full wind and superimposed loading acting at the same time.

Thus, corresponding to the load factors implied by BS 449, the plastic designer would carry out a design under dead plus superimposed load at a load factor of 1·75, and a second design under dead plus superimposed plus wind load at a load factor of 1·40; naturally, the more severe of these two conditions would govern the final design. These values of load factor appear to be safe for the type of structure for which BS 449 was devised; for other, different, structures, there might be a case for either increasing or decreasing the load factor used for design, depending on the designer's estimate of the uncertainty of the design conditions.

The idea of a single load factor applied to all the loads acting on a particular structure is rather crude, but will serve in this present volume to illustrate the basic approach of plastic theory to the design of steel frames. More sophisticated practical design methods will take account of the fact that, for example, dead loads are usually known with far greater accuracy than say wind loads. In such a case the loading pattern to be used for design can be assembled incorporating various load factors for the different classes of loading.

EXAMPLES

1.1 For each of the following cross-sections, calculate the value of the plastic modulus.

(a) A square of side a bent about a diagonal.
(*Ans.* $a^3/3\sqrt{2}$.)

(b) A solid circular section, diameter d.
(*Ans.* $d^3/6$.)

(c) A thin-walled circular tube, *mean* diameter d, wall thickness t.
(*Ans.* $\frac{1}{6}(d+t)^3 - \frac{1}{6}(d-t)^3 \simeq td^2$.)

(d) An I-section, overall depth 300 mm, flanges 200 × 20 mm, web thickness 10 mm, bent about (i) strong axis, and (ii) weak axis. (Assume three rectangles.)
(*Ans.* (i) 1289 000 mm³; (ii) 406 500 mm³.)

(e) The channel section shown, constrained to bend in the xx plane (i.e. bent about the weak axis).
(*Ans.* 240 000 mm³.)

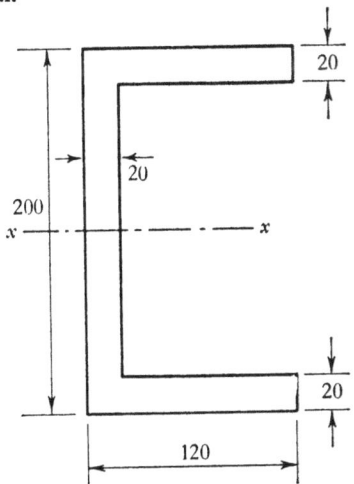

37

(*f*) The channel section shown (i) bent in the *xx* plane, and (ii) bent about the *xx* axis.

(*Ans.* (i) 390 000 mm^3; (ii) 840 000 mm^3.)

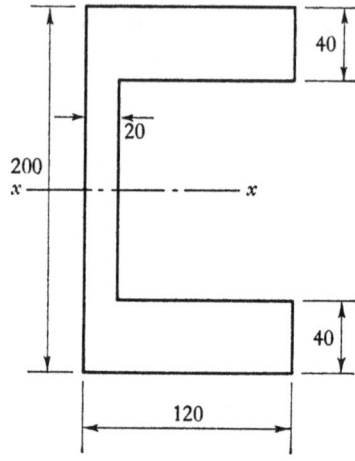

(*g*) An isosceles triangle of height *h* and base *a*, bent about an axis parallel to the base.

$$\left(Ans.\ ah^2\left(\frac{\sqrt{2}-1}{3\sqrt{2}}\right).\right)$$

(*h*) An I-section, made up of three rectangles, overall depth *h*, flange width *b* and thickness t_1, web thickness t_2, having in addition on each flange a plate of width *w* and thickness t_3, bent (i) about the strong axis, and (ii) about the weak axis.

$$(Ans.\ (\text{i})\ bt_1(h-t_1)+wt_3(h+t_3)$$
$$+\tfrac{1}{4}t_2(h-2t_1)^2;$$
$$(\text{ii})\ \tfrac{1}{2}b^2t_1+\tfrac{1}{2}w^2t_3+\tfrac{1}{4}t_2^2(h-2t_1).)$$

(*j*) An I-section, made up of three rectangles, overall depth 300 mm, flanges 200 × 20 mm, web thickness 10 mm, having on the upper flange only an additional plate 250 × 20 mm, bent about the strong axis.

(*Ans.* 1 600 800 mm^3.)

(*k*) A T-section, width 100 mm, depth 120 mm, composed of two equal rectangles 100 × 20 mm, bent about the strong axis. (*Ans.* 120 000 mm^3.)

(*l*) A T-section, of uniform thickness 10 mm, width 150 mm and depth 150 mm. (*Ans.* 105 483 mm^3.)

(*m*) The castellated beam shown made from the I-section of example (*d*).

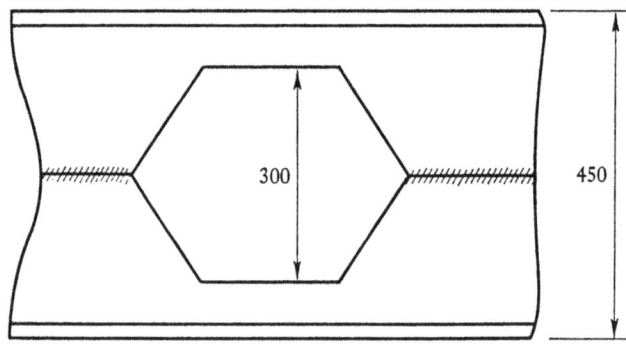

(*Ans.* (i) 2 140 250 or 1 915 250 mm^3; (ii) 410 250 or 402 750 mm^3.)

(*n*) The gantry girder shown made up from a 36 × 12 UB 289 kg, a channel 17 × 4 × 65·48 kg, and a bottom flange plate 360 × 15 mm. (The channel has area 8349 mm² and web thickness 12·2 mm; its *yy* axis is 23·2 mm from the back face).

(*Ans.* 18 770 000 mm³.)

(*o*) A box section made by welding toe to toe two 71 × 4 × 65·48 kg channel bent about the axis through the welds. (Section details in example (*n*); each channel has flange width 102 mm.)

(*Ans.* 1 316 000 mm³.)

(*p*) The square shaft shown, 8*d* × 8*d*, having a keyway 4*d* × 2*d*, bent about the *xx* axis.

(*Ans.* 128*d*³ − 26*d*³ = 102*d*³.)

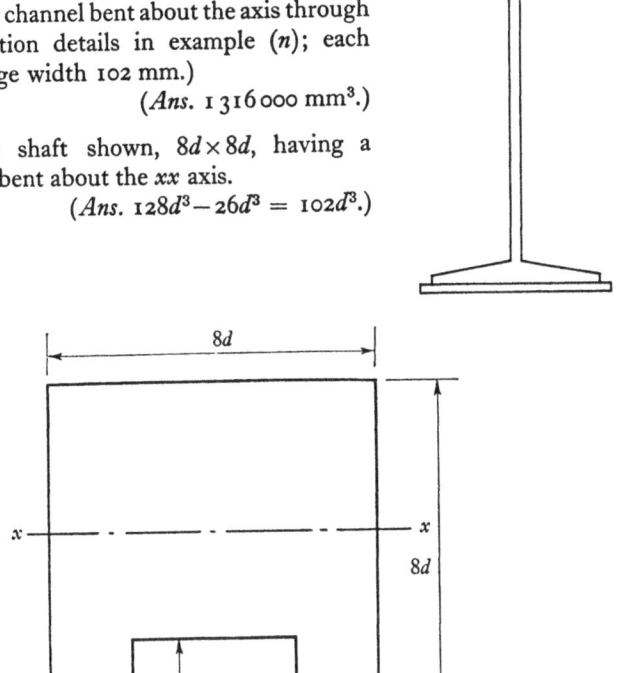

1.2 Design an I-section, 500 mm deep and 200 mm wide, made up of rectangles, which will have a plastic modulus of 2 449 000 mm³ about the strong axis and 411 500 mm³ about the weak axis.

(*Ans.* Flange thickness 20 mm; web thickness 10 mm.)

1.3 A steel T-beam, 160 mm wide and 240 mm deep, flange thickness 20 mm, and unknown web thickness, was simply supported with the flange horizontal over a span of 4 m. The beam collapsed when a uniformly distributed load of 250 kN was applied. A compression test on a short length of the T-section gave a squash load of 2000 kN. What was the web thickness?

(*Ans.* 16·3 mm.)

1.4 A beam of uniform section throughout is simply supported over a span of 10 m. It carries two transverse loads, one of 100 kN at a distance 3 m from the left-hand support, and one of 200 kN at a distance 1 m from the right-hand support The load factor of the beam is 2 under this load system; what is its full plastic moment? (*Ans.* 540 kNm.)

1.5 What additional transverse load acting at midspan would need to be added to the beam of example 1.4 just to bring about collapse?
(*Ans.* 116 kN.)

1.6 A straight steel beam, 1 m long, is simply supported at its ends. It is of rectangular section 50 mm deep throughout, but with a width tapering uniformly from 50 mm at the centre to 25 mm at each end. When a transverse load of 30 kN is applied 0·3 m from one end the beam collapses. What is the yield stress of the steel? (*Ans.* 252 N/mm².)

1.7 A straight steel beam, 1 m long, is simply supported at its ends. It is of rectangular section 50 mm deep throughout, but with a width tapering from 75 mm at the centre to 25 mm at each end. If the yield stress of the steel is 240 N/mm², what uniformly distributed load will bring about collapse?
(*Ans.* 78·5 kN.)

1.8 A beam of rectangular but varying cross-section, symmetrical about the centre of length, is simply supported at its ends. The beam has a uniform depth, but tapers uniformly from a width b at the centre towards either end. Collapse occurs under uniformly distributed load by the formation of hinges at quarter span. What is the width of the beam at its ends? (*Ans.* $\frac{1}{3}b$.)

1.9 If the full plastic moment of the beam of example 1.8 has the value of $5M_0$ at the centre, and the span is l, find the value of the collapse load.
(*Ans.* $32M_0/l$.)

1.10 A beam of rectangular cross-section, 40 mm wide and 60 mm deep, is simply supported over a span of 0·75 m; the steel has a yield stress 250 N/mm². The beam carries a central point load of 40 kN. What is the length of the plastic zone, and the depth of the elastic core at mid-span?
(*Ans.* 150 mm, 42·4 mm.)

1.11 What is the central deflexion of the beam in example 1.10? (Take E as 205 GN/m².) (*Ans.* 2·43 mm.)

1.12 A beam of rectangular cross-section is simply supported and subjected to a concentrated load one-third of the span from the support. Plot the shape of the plastic zone at collapse of the beam.

1.13 Plot the shape of the plastic zone at collapse for a beam with the uniform cross-section of example 1.1(*d*) when simply supported with flanges horizontal, and subjected to a central load.

1.14 Repeat example 1.13 for a uniformly distributed load.

1.15 Show that the M/κ relationship for a beam of rectangular section in the plastic range is

$$\frac{M}{M_y} = \frac{3}{2}\left\{1 - \frac{1}{3}\left(\frac{\kappa_y}{\kappa}\right)^2\right\}.$$

1.16 Show that the M/κ relationship for a beam of square cross-section bent about a diagonal is

$$\frac{M}{M_y} = 2\left\{1 - \left(\frac{\kappa_y}{\kappa}\right)^2 + \frac{1}{2}\left(\frac{\kappa_y}{\kappa}\right)^3\right\}.$$

1.17 Show that the M/κ relationship for an I-section, flange width b, overall depth $2d$, flange thickness, T, web thickness t is

$$\frac{M}{M_y} = \frac{\kappa}{\kappa_y}\left(1 - \frac{2bd^2}{3Z_e}\right) + \frac{bd^2}{Z_e}\left\{1 - \frac{1}{3}\left(\frac{\kappa_y}{\kappa}\right)^2\right\}$$

for the range

$$1 \leqslant \frac{\kappa}{\kappa_y} \leqslant \frac{d}{d-T},$$

and

$$\frac{M}{M_y} = \frac{Z_p}{Z_e} - \frac{td^2}{3Z_e}\left(\frac{\kappa_y}{\kappa}\right)^2$$

for the range

$$\frac{\kappa}{\kappa_y} \geqslant \frac{d}{d-T}.$$

1.18 Derive equation (1.8) from equations (1.4) and (1.7).

1.19 A beam having the section described in example 1.1(d) is made of steel with a yield stress 250 N/mm². Calculate the percentage reductions in the full plastic moments about the principal axes due to an axial load of

(*a*) 25 N/mm², and (*b*) 125 N/mm².

(*Ans.* (*a*) Strong 2·18%; Weak 0·23%. (*b*) Strong 41·0%; Weak 12·8%.)

1.20 Flange plates 250 × 25 mm are added to the flanges of the beam in example 1.19. Find the increase in the full plastic moment about the strong axis. Calculate the reduction in full plastic moment due to an axial load of 1000 kN. (*Ans.* 508 kNm, 88 kNm.)

1.21 If the flange plates in example 1.20 had been only 12 mm thick, by how much would the full plastic moment have been reduced by the application of the axial load of 1000 kN? What would be the reduction if the axial load were doubled? (*Ans.* 88 kNm, 227 kNm.)

1.22 What is the effect of the addition of flange plates to an I-section on the percentage reduction in full plastic moment due to the application of an axial load of given intensity? Illustrate by comparing the percentage reduction of

the full plastic moment about the strong axis due to an axial load of 125 N/mm²
for the section of example 1.19 with those for the plated sections of
examples 1.20 and 1.21.   (*Ans.* 41·0%, 43·1%, 42·8%.)

1.23 A beam having the section described in example 1.1(*a*) is made of
steel with a yield stress σ_0. Calculate the percentage reduction in full plastic
moment due to a mean axial stress of $\frac{1}{2}\sigma_0$.   (*Ans.* $50(\sqrt{2}-1) = 20\cdot7\%$.)

1.24 A 12×5 UB 48 kg Universal Beam is made of steel having a yield
stress of 250 N/mm². Calculate the percentage reduction in full plastic
moment about the strong axis for a mean axial stress of (*a*) 100 N/mm² and
(*b*) 200 N/mm². (Use equations (1.29) and (1.30).)   (*Ans.* 23·6%, 73·7%.)

1.25 Determine the reduction in the plastic modulus about the strong axis
for the I-section of example 1.1(*d*) due to the application of a shear force
of 250 kN in the plane of the web when the steel has a yield stress in tension
of (*a*) 250 N/mm², and (*b*) 350 N/mm². Use the Mises criterion.
                    (*Ans.* (*a*) 43 000 mm³; (*b*) 20 000 mm³.)

1.26 A $12 \times 6\frac{1}{2}$ UB 40 kg is built in to a support at one end with its flanges
horizontal. Vertical loads of 50 kN and 150 kN are applied 0·5 m and 0·7 m
respectively from this end. Determine the load factor against collapse
(*a*) neglecting the effect of shear, and (*b*) taking shear into account. The yield
stress in tension of the steel is 250 N/mm²; use the Mises criterion.
                    (*Ans.* 1·20, 1·08.)

1.27 Show that for an I-section subjected to a shear force F the full plastic
moment about the strong axis is

$$M_f + \sqrt{\left\{M_w\left(M_w - \frac{3}{4}\frac{F^2}{t\sigma_0}\right)\right\}},$$

where M_f and M_w are the contributions of the flanges and the web to the full
plastic moment in the absence of shear force, t is the thickness of the web, and
σ_0 is the yield stress in tension.

1.28 A horizontal centilever beam, length 3 m, is of uniform cross-section
having full plastic moment 60 kNm. A downward vertical load of 10 kN is
applied 0·6 m from the support. What additional upward load could be
applied at the free end before collapse occurred? What could this load be if
the downward vertical load were increased to 30 kN? Neglect the effect of
shear.   (*Ans.* 22 kN, 25 kN.)

1.29 A horizontal centilever beam of uniform cross-section and length l
collapses under a transverse end load W. If an upward vertical load nW is
applied at a distance ml from the support, what vertical load can now be
carried at the end? Neglect the effect of shear.
(*Ans.* $nm \leqslant 1$. Upward load: $W(1-mn)$
                    Downward load: The smaller of $W(1+mn)$ and $W/(1-m)$.)

1.30 The only member available to support a uniformly distributed load of 260 kN over a span of 8 m is a 15×6 UB 67 kg in steel having a yield stress 250 N/mm². Determine the cross-section and length of symmetrical flange plates (of the same steel) which must be added to provide a load factor of 1·75. The ends of the member are simply supported; neglect the effect of shear.

(*Ans.* 180 × 8 mm; 4·5 m.)

1.31 As originally designed a balcony, 5 m wide, was to be supported by $10 \times 5\frac{3}{4}$ UB 31 kg cantilevers of mild steel ($\sigma_0 = 250$ N/mm²) spaced at 2 m centres so that each beam carried a uniformly distributed load of 20 kN. It was then decided to add a heavy balustrade which would apply an additional load of 10 kN to the free end of each beam. Find the load factor of the beams as originally designed and the dimensions of the symmetrical flange plates which must be added to provide a load factor of 2 when the balustrade is in position. Neglect the effect of shear. (*Ans.* 1·98; 160 × 10 mm.)

1.32 Select a Universal Beam ($\sigma_0 = 250$ N/mm²) to support the loads shown with a load factor of 1·75. Neglect the effect of shear.

(*Ans.* $Z_p = 2\,730\,000$ mm³, say $21 \times 8\frac{1}{4}$ UB 109 kg.)

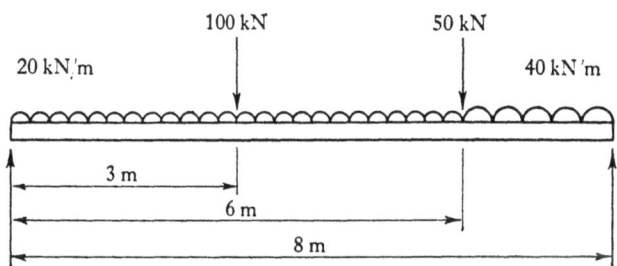

1.33 Design a simply supported plate girder of uniform section, with a web depth of 1·4 m, to carry a uniformly distributed load of 4000 kN over a span of 15 m with a load factor of 1·75 ($\sigma_0 = 250$ N/mm²). It may be assumed that the girder is adequately supported laterally.

(*Ans.* Web: 18 mm; flanges 600 × 50 mm.)

2

SIMPLE BEAMS AND FRAMES

2.1 Collapse of a redundant beam

The ease and power of plastic methods become apparent when they are applied to the structural problem proper, that is, to the analysis and design of redundant structures. In the discussion which follows, it is convenient to make a further idealization in the moment–curvature relationship for the bending of a beam. In fig. 2.1 this relationship has

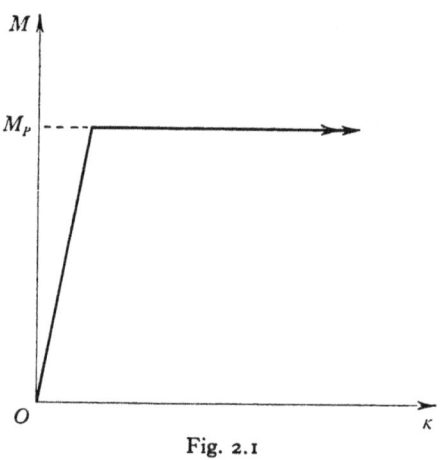

Fig. 2.1

been simplified to admit only of elastic or of perfectly plastic states (cf. fig. 1.6); that is, the shape factor of the cross-section is taken as unity. Thus a bending moment M will produce entirely elastic curvatures if its numerical value is less than M_p; alternatively, for $M = \pm M_p$, a plastic hinge is formed which can undergo indefinite rotation. This idealization, while simplifying the numerical work, does not restrict the validity of the arguments; it will be seen that the final collapse load of a frame depends only on the value of M_p and not on the complete moment–curvature relationship.

Suppose the encastré beam of fig. 2.2(a), which has its ends so fixed that they can neither rotate nor deflect, has a uniform cross-section with the idealized moment–curvature characteristic, and is subjected to a slowly increasing load W until collapse occurs. If W is small, the bending moments in the beam will be elastic, and the distribution will be as shown in fig. 2.2(b). It is of interest to enumerate the steps in the elastic analysis which lead to this elastic bending moment distribution.

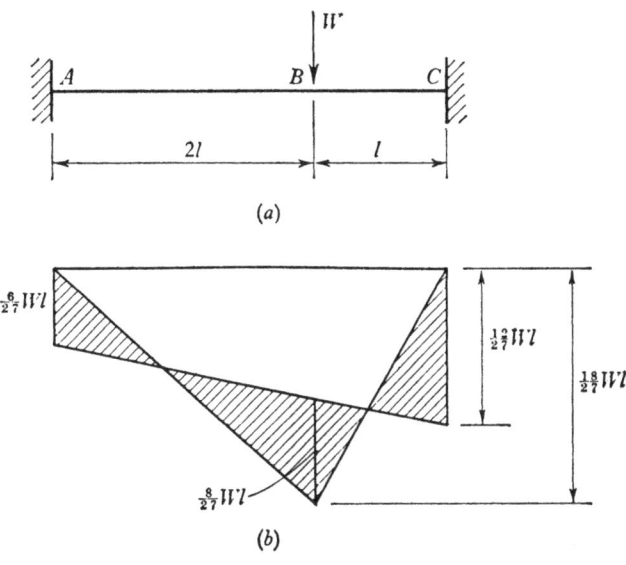

(a)

(b)

Fig. 2.2

In conventional terms, the beam has two redundancies, that is, the use of equilibrium equations alone will determine the bending moments in the beam only in terms of two unknown quantities. There is a fairly wide choice even for this simple problem as to which quantities shall be labelled as the redundancies; in fig. 2.3, the bending moments M_A and M_C acting at the ends of the beam are shown as the two unknowns. The beam of fig. 2.3 is in equilibrium, and no further information can be obtained by the use of simple statics, that is, by resolving forces and taking moments.

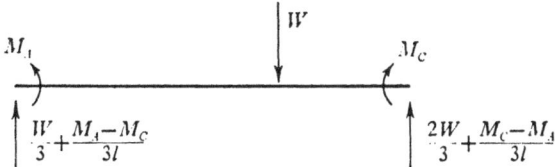

Fig. 2.3

It is because equilibrium equations alone will not suffice to determine the elastic solution that the problem is called statically indeterminate, or redundant. In order to obtain the solution, the deformation of the beam must be considered, and use must be made of the fact that the ends can neither rotate nor deflect. The deformation of the beam is governed by the moment–curvature relationship; if the beam is wholly elastic then, with the usual notation,

$$M = EI\kappa = EI\frac{\mathrm{d}^2 y}{\mathrm{d}x^2}. \tag{2.1}$$

If the coordinate origin is taken at the left-hand end of the beam in fig. 2.3, then, using Macaulay's notation, the differential equation of elastic bending of the beam may be written

$$EI\frac{\mathrm{d}^2 y}{\mathrm{d}x^2} = M_A - \left(\frac{W}{3} + \frac{M_A - M_C}{3l}\right)x + W[x - 2l], \tag{2.2}$$

where the term in square brackets is ignored if it is negative. Equation (2.2) may be integrated to give

$$EI\frac{\mathrm{d}y}{\mathrm{d}x} = M_A x - \left(\frac{W}{3} + \frac{M_A - M_C}{3l}\right)\frac{x^2}{2} + \tfrac{1}{2}W[x - 2l]^2 + A, \tag{2.3}$$

and again to give

$$EIy = \tfrac{1}{2}M_A x^2 - \left(\frac{W}{3} + \frac{M_A - M_C}{3l}\right)\frac{x^3}{6} + \frac{W}{6}[x - 2l]^3 + Ax + B. \tag{2.4}$$

Equations (2.3) and (2.4) expressing the slope and deflexion respectively of the beam at any point are written in terms of four unknown quantities, namely the two redundancies M_A and M_C and the two constants of integration A and B. These four unknowns may be found from the four boundary conditions that the slope and deflexion of the beam at both ends are zero. It will be seen that the constants A and B are both zero, and that M_A and M_C have the values noted in fig. 2.2(b), $\tfrac{2}{9}Wl$ and $\tfrac{4}{9}Wl$.

The complete solution of this problem has thus required, in addition to the central statement of equilibrium, a knowledge of the material properties (as expressed by the moment–curvature relationship) and some simple compatibility statements (boundary conditions on the displacements). The moment–curvature relationship need not be linear, as assumed in equation (2.1); it is only so when attention is focused on the elastic problem. Schematically, the equations used were examples of the three master equations of the theory of structures applied to frames, namely those of

COMPATIBILITY (Displacement conditions),

EQUILIBRIUM (Statics),

MOMENT–CURVATURE RELATION $(M = EI\kappa)$.

$$\left.\begin{array}{l} \\ \\ \\ \end{array}\right\} \quad (2.5)$$

The elastic bending-moment diagram of fig. 2.2(b) will be valid so long as the largest bending moment $(M_C = \frac{12}{27}Wl)$ is less than M_p. As the load W is slowly increased, therefore, a plastic hinge will first form (at the load $W = 9M_p/4l$) at the end C of the beam. Further slight increase in the value of W will cause the hinge to rotate but cannot, from fig. 2.1, result in any further increase in the value of M_C; the bending moment at C must remain at the value M_p, fig. 2.4(a). In order to calculate the new distribution of bending moments, a new analysis must be made, but it will be seen that this is easier than that for the original problem. The value of the bending moment is now known at the end C of the beam to be equal to M_p, so that only a single redundancy remains, the value of M_A.

Equation (2.2) can be rewritten with $M_C = M_p$, so that equations (2.3) and (2.4) for the deflexion and slope of the beam now involve only three unknowns, namely the value of M_A and the two constants of integration. As before, the slope and deflexion at the end A of the beam must be zero, and the deflexion at C must also be zero; these are the three required conditions necessary to complete the solution sketched in fig. 2.4(b). The original fourth condition, that the slope at C is zero, no longer holds, since the plastic hinge is rotating; the reduction in degree of redundancy by one is exactly balanced by the loss of a boundary condition on the displacements.

When the value of W is increased to $81M_p/28l$, a second plastic hinge forms under the load at B, and further increase in W will be accompanied by rotation at both hinges B and C, fig. 2.5(a). There has been a further reduction in the degree of redundancy, from one to zero, and the beam is now statically determinate; the value of M_A can be found immediately, and the bending moment diagram, fig. 2.5(b), can be drawn without the need of writing the differential equation of bending. The compatibility condition that has been destroyed by the formation of the hinge at B is an internal condition of continuity; the slope of the beam is no longer continuous at B.

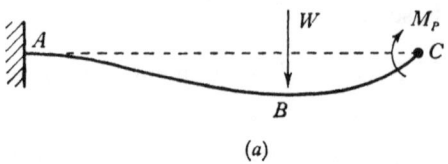

(a)

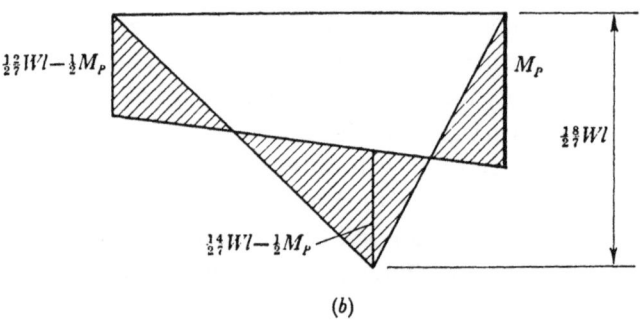

$\frac{12}{27}Wl - \frac{1}{2}M_P$

M_P

$\frac{13}{27}Wl$

$\frac{14}{27}Wl - \frac{1}{2}M_P$

(b)

Fig. 2.4

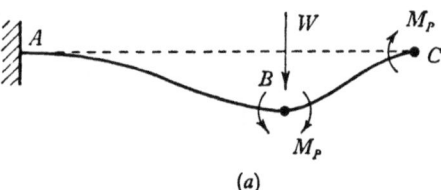

(a)

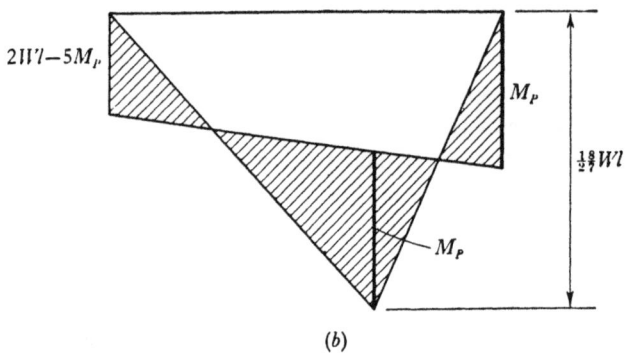

$2Wl - 5M_P$

M_P

$\frac{13}{27}Wl$

M_P

(b)

Fig. 2.5

When the value of W is increased to $3M_p/l$, a final hinge is formed at A. The two previous hinges had already turned the originally re-dundant beam into a statically determinate structure; the formation of the third hinge turns the beam into a *mechanism*. In general, each hinge as it forms reduces the degree of redundancy; the final hinge transforms the beam into a mechanism of collapse.

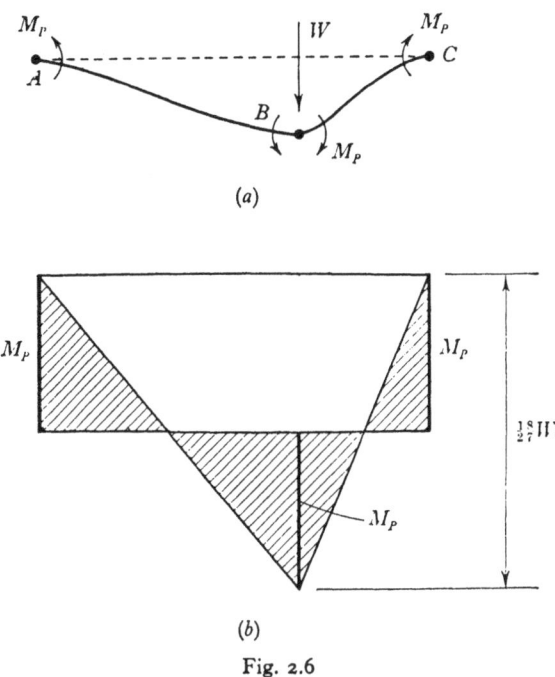

(a)

(b)

Fig. 2.6

For each of the beams in figs. 2.2, 2.4 and 2.5, it is relatively simple to compute deflexions from the differential equation of bending. Plotting the value of the load W against the deflexion of the loading point B, the curve of fig. 2.7 is obtained. The sharply differentiated portions of the curve result from the idealized elastic-perfectly plastic moment–curva-ture relationship, fig. 2.1, that was assumed. If the more realistic relationship of fig. 1.6 had been used, the load-deflexion curve would have had rounded corners, shown dotted in fig. 2.7, rather than the abrupt changes of slope; otherwise, the same essential features of the load-deflexion curve would have been reproduced.

In particular, the final collapse load $W = 3M_p/l$ depends on the value of M_p alone; while the precise form of the load-deflexion curve is affected

49

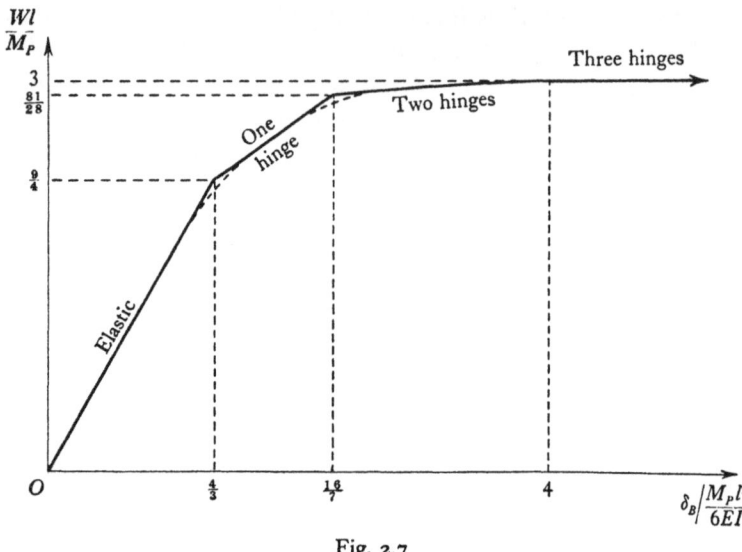

Fig. 2.7

by the form of the moment–curvature relationship, it is only the value of the full plastic moment which is used in writing the collapse equation.

Further, it is clear that the final value of the collapse load does not depend on the order of formation of the hinges. Thus the tracing of the complete loading history of a beam or frame is not required for the purposes of simple plastic analysis (or design). For the elementary problem discussed above, the limiting state of fig. 2.6 can be determined by inspection, as will be clear from fig. 2.9 below. Such a straightforward examination is of course not possible for more complex frames, but the essential point remains that the frame needs to be examined only in its collapse state, and that a step-by-step analysis is not required.

A plastic design process is therefore likely to be much simpler than the corresponding elastic process, if the features of this simple example carry through to more complex frames. The plastic design process will certainly, and perhaps curiously, also bear a closer relationship to the actual behaviour of the frame. It was seen that the elastic solution required the simultaneous deployment of the three master statements of structural theory, equations (2.5). Of these three statements, the compatibility conditions entered into the problem as certain boundary conditions on the deformations (slopes and deflexions both zero at both ends of the beam). Thus in equations (2.3) and (2.4) both y and dy/dx were set equal to zero for $x = 0$, $3l$.

Such boundary conditions are notoriously difficult to satisfy in practice, and an examination of the elastic equations (e.g. (2.3) and (2.4)) shows that very small settlements or rotations of supposedly fixed supports can make very large differences to the computed values of the bending moments. It has in fact been observed from tests on real frames that actual bending moments bear almost no relation to those computed, and the elastic design process appears from this point of view to be very irrational.

However, even if small settlements, and so on, lead to large calculated and observed changes in the values of bending moment, it seems intuitively 'obvious' that accidental imperfections of this kind will have little effect on the real *strength* of a frame. Intuition is supported here by the results of plastic analysis. In the example of the fixed-ended beam, the final collapse state involved the formation of hinges, with consequent rotation, at the two ends of the beam; the compatibility conditions ($dy/dx = 0$ at $x = 0, 3l$) were destroyed at precisely the locations most suspect from the practical point of view. Further, the process of the formation of the hinges led to a statically determinate set of bending moments in the beam, which could be calculated without reference to compatibility conditions at all.

Thus although the conditions $dy/dx = 0$ at $x = 0, 3l$ no longer hold in equation (2.3), there is in fact no need to write this equation, since the values of M_A and M_C are known since two hinges have formed. Similarly, equation (2.4) governing the deflexions of the beam need not be written if only the equation governing collapse of the beam is required. The final bending moment diagram of fig. 2.6(*b*) is completely independent of any practical imperfections involving rotation or settlement of the two ends of the beam (it is always assumed that deformations are not so gross as to upset the overall geometry of the structure). The collapse load $W = 3M_p/l$ will be recorded independently of the precise way in which the beam is connected to its supports, providing only that the connexions are strong enough so that the full plastic moment M_p can be developed.

The bending-moment diagrams of figs. 2.2, 2.4, 2.5 and 2.6 are all similar in form; they are, in fact, special cases of the general diagrams of fig. 2.8. The construction of fig. 2.8 may be seen from an examination of fig. 2.9. Under the application of the external load W to the encastré beam of fig. 2.9(*a*), bending moments M_A and M_C are induced at the ends of the beam, so that the net bending-moment diagram is as shown in fig. 2.9(*b*). This bending-moment diagram can be built up by imagin-ing the external load W to be applied to a statically determinate beam such as the simply-supported beam of fig. 2.9(*c*), giving rise to the 'free' moment diagram of fig. 2.9(*d*). To obtain the true bending-moment diagram the effect of the end restraining moments M_A and M_C (fig. 2.9(*e*)), giving rise to the 'reactant' moment diagram of fig. 2.9(*f*),

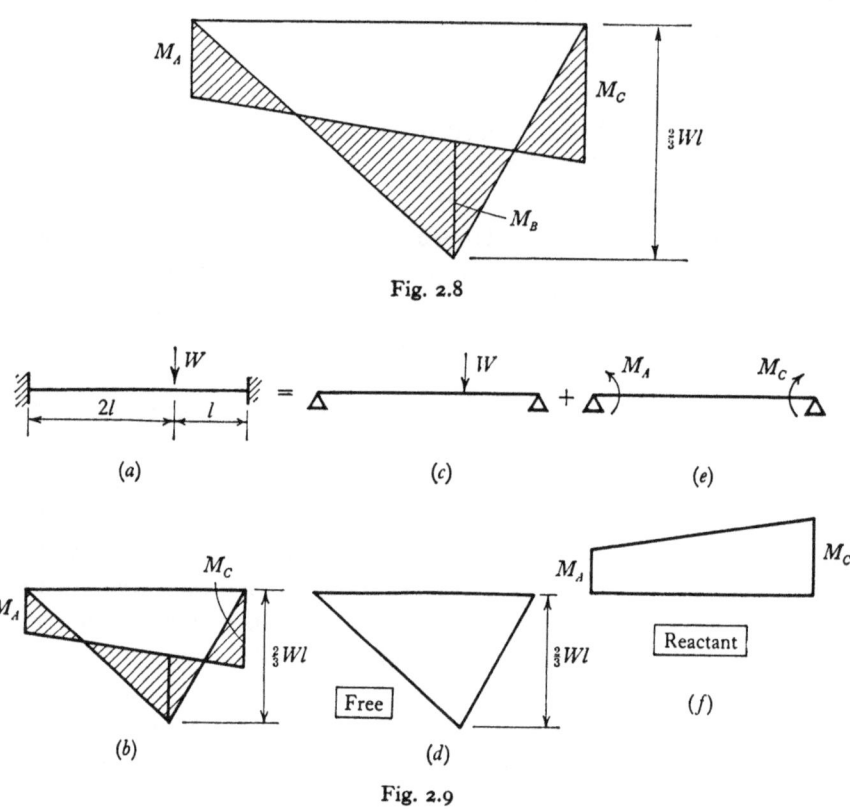

Fig. 2.8

Fig. 2.9

must be added. The general bending-moment diagram of fig. 2.8 results from the superimposition of these free and reactant moments.

Bending-moment diagrams can always be constructed in this way, by the addition of free and reactant components, irrespective of whether the material has a linear elastic characteristic or is partially or com-pletely plastic. The equilibrium equations giving the values of the bend-ing moments are themselves always linear, as exemplified by equation (2.2), in which the bending moment is a linear function of W, M_A and

M_C. The graphical addition carried out in fig. 2.9 is no more than the analytical addition which occurs in the equation for the bending moments.

Figure 2.8 is, in fact, the completely general bending-moment diagram for the fixed-ended beam carrying an eccentric point load. The triangle of height $\frac{2}{3}Wl$ is fixed from the value and location of the external load; this is the free bending-moment diagram. On this diagram is superimposed a straight reactant line resulting from the end moments; the shape of the reactant line does not depend on the external loading, but is basically a property of the geometry of the frame.

The essential structural problem, elastic or plastic, consists in positioning the reactant line, i.e. in calculating the values of the redundant quantities M_A and M_C. In fig. 2.8 it will be seen that some relationship must exist between the three values M_A, M_B and M_C; by simple proportion, it will be found that

$$M_A + 3M_B + 2M_C = 2Wl. \tag{2.6}$$

Thus, if any two of the three bending moments are known, the third can be calculated, but there is no way from statics alone of finding those two values. Figure 2.8 is a graphical representation of the application of statics to the beam under the action of the load W, and the beam is twice indeterminate.

It was seen that the elastic solution required the use of the moment–curvature relationship ($M = EI\kappa$) and also of certain compatibility conditions, in addition to the central statement of equilibrium, equations (2.5). Similar extra conditions are required for the plastic solution, but these are much simpler than the corresponding elastic statements. These extra plastic conditions can be deduced by inspection of fig. 2.8. It was noted that the elastic compatibility conditions were destroyed by the formation of plastic hinges, and that the last hinge transformed the statically determinate beam into a mechanism. Thus the deformation statement required by simple plastic theory is that sufficient plastic hinges occur to form a mechanism of collapse. In fig. 2.6, the reactant line has been positioned so that $M_A = M_B = M_C = M_p$; directly from the diagram, or from equation (2.6), the collapse condition is given by $2M_p = \frac{2}{3}Wl$.

It is clear by inspection of fig. 2.6 that no bending moment exceeds the value M_p, but this will not be so obvious in more complex examples. In general it must be verified that the *yield condition*, as it is usually called, is satisfied, that is, that the bending moments everywhere in a collapsing beam or frame do not exceed the local value of M_p. Thus the three master statements of plastic analysis and design, corresponding to the three elastic statements of (2.5), become those of

$$\left.\begin{array}{l} \text{MECHANISM,} \\ \text{EQUILIBRIUM,} \\ \text{YIELD } (|M| \leqslant M_p). \end{array}\right\} \qquad (2.7)$$

2.2 Graphical analysis and design

In fig. 2.10 are given some simple examples of plastic collapse analysis for beams of uniform section throughout. Quite unlike the corresponding elastic problems, the collapse bending moment-diagrams can be sketched immediately. For example, the fixed-ended beam carrying a

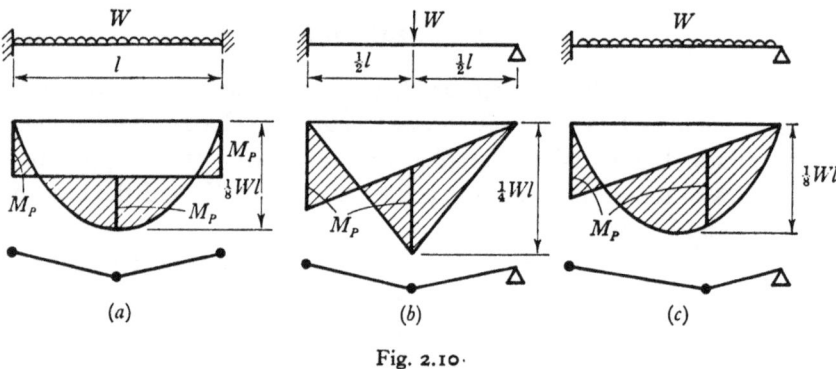

Fig. 2.10.

uniformly distributed load will form a symmetrical collapse mechanism with three hinges; it will be seen from fig. 2.10(a) that $2M_p = Wl/8$. Similarly, inspection of the collapse bending-moment diagram for the propped cantilever carrying a central point load, fig. 2.10(b), leads to the equation $M_p + \frac{1}{2}M_p = Wl/4$, or $M_p = \frac{2}{3}(Wl/4) = Wl/6$. This and the previous equation can be interpreted as relating either to design or to analysis. If, for example, the value of W is taken to incorporate a suitable load factor, then the value of M_p given by the collapse equation represents the design of a beam of uniform section that will just carry the load. Alternatively, for a beam of given cross-section and hence a known value of M_p, the equation may be solved to give the value of the load W which will just cause collapse.

The diagram for the propped cantilever carrying a uniformly distributed load, fig. 2.10(c), is again a proper combination of free and reactant bending moments which satisfies each of the three statements (2.7). The parabola, of height $Wl/8$, is the same parabola as that of fig. 2.10(a); the reactant line has been superimposed subject to the condition that the bending moment is zero at the right-hand (pinned) end of the beam. By inspection, the value of M_p at collapse is about $\frac{2}{3}(Wl/8)$; however, it is clear that the sagging hinge forms at a cross-section somewhat off-centre. The analytical determination of the exact value of M_p is left until later in the chapter, but the value may be found quickly and easily by drawing to confirm the collapse equation

$$M_p = 0.686(Wl/8). \tag{2.8}$$

For the time being, the complication of distributed loads will be avoided, and the next few examples will involve concentrated loads only. (Further examples involving distributed loads will be found at the end of the chapter.)

The two-span beam of fig. 2.11 carries unequal central point loads, and it is clear that if the beam is of uniform section throughout, then the left-hand span will govern the design. The free bending-moment diagram in fig. 2.11(b) has been drawn by setting the moment at the central support equal to zero, i.e. by considering the beam as two simply-supported spans. The reactant line due to the bending moment at the central support has been superimposed on the free bending-moment diagram in fig. 2.11(c), and has been arranged to correspond to collapse of the left-hand span, fig. 2.11(d). The mechanism and equilibrium conditions of (2.7) are therefore satisfied, since fig. 2.11(c) is a proper combination of free and reactant bending moments and corresponds to a mechanism of collapse, with $M_p = 6$. The bending moment under the right-hand load has value 3 units, so that the yield condition of (2.7) is also satisfied. The simultaneous satisfaction of the three master conditions means that the correct solution has been obtained.

This last statement will be amplified in the next chapter; it is of interest here to examine the incorrect solution of fig. 2.12. It has been assumed that the right-hand span collapses, leading at once to the value $M_p = 4$. It is apparent that the yield condition is violated under the left-hand load, where the bending moment has the value 7 units. Thus, although the conditions of mechanism and equilibrium are satisfied by the solution of fig. 2.12, this solution is physically impossible in

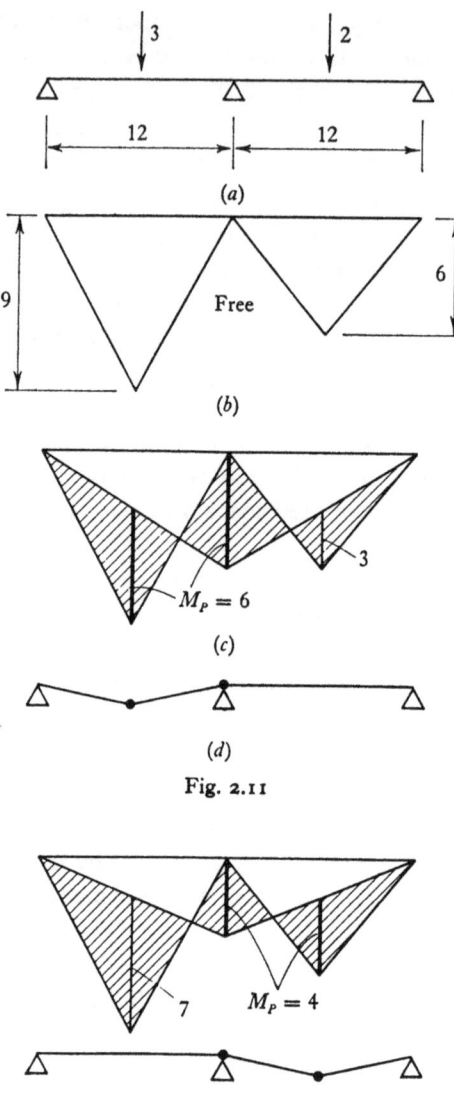

Fig. 2.11

Fig. 2.12

that a continuous beam of full plastic moment 4 units is required to carry a bending moment of 7 units.

A wrong assumption of a collapse mechanism has led to a wrong, and, moreover, unsafe solution to the problem. The correct solution, fig. 2.11, required provision of a beam having a full plastic moment of at least $M_p = 6$, whereas the incorrect solution of fig. 2.12 requires M_p to be only 4 units. This 'unsafe' property of solutions corresponding to incorrect mechanisms of collapse is general, and will be commented upon in the next chapter. For the moment, the statement may be made that any solution derived from an arbitrarily assumed mechanism is unsafe, or at best correct if the mechanism happens to be correct.

This statement indicates a possible approach to the problem of plastic design of complex frames. All possible mechanisms of collapse could be examined individually, and the corresponding values of full plastic moment computed; the final design would be based on the mechanism which gave the largest value of M_p. Such an approach is possible for simple frames, but becomes unworkable, because of the large number of mechanisms, for a frame of any complexity. For continuous beams, however, this direct attack is convenient for both analysis and design.

A further use may be made of the wrong solution of fig. 2.12. For a beam of uniform cross-section, the bending-moment diagram corresponds neither to the elastic solution nor to the plastic collapse state; however, it is a proper combination of correctly drawn free and reactant bending-moment diagrams and hence is a *possible* equilibrium state of the beam, satisfying at least the second of the conditions (2.7). It seems reasonable (and, again, this can be proved; the general theorem is given in the next chapter) that if a beam of uniform section were provided having $M_p = 7$, corresponding to the largest bending moment in fig. 2.12, then such a design would be safe. A solution, in fact, which is in equilibrium and which satisfies the yield condition, but which does not correspond to a mechanism of collapse, is a *safe* solution. Thus both a safe and an unsafe estimate of the value of M_p can be derived from fig. 2.12:

$$7 \geqslant M_p \geqslant 4. \tag{2.9}$$

(It may be noted that the correct value $M_p = 6$ falls within this range.) These bounds are too wide for design purposes, but not, perhaps, much too wide, since M_p is given from inequalities (2.9) as 5·5 to within $\pm 27\%$. However, other, more complex, frames will lead to solutions giving much closer bounds which would certainly be within practical tolerances. These ideas of bounds to the required quantities will be discussed more fully later.

It has been assumed, in discussing the problem of figs. 2.11 and 2.12, that the same cross-section was used for both spans of the beam. If two different sections were used, one for each span, then the bending-moment distribution of fig. 2.12 clearly corresponds to the minimum possible section ($M_p = 4$) that could be used for the right-hand span. If now a section having $M_p = 7$ were provided in the left-hand span, a perfectly good design of the beam system would result, fig. 2.13. Thus *direct design* is possible by plastic theory in a way which is completely denied by the use of a conventional elastic approach; moreover, a designer can investigate quickly and easily the effect of altering the proportions of the members of a frame.

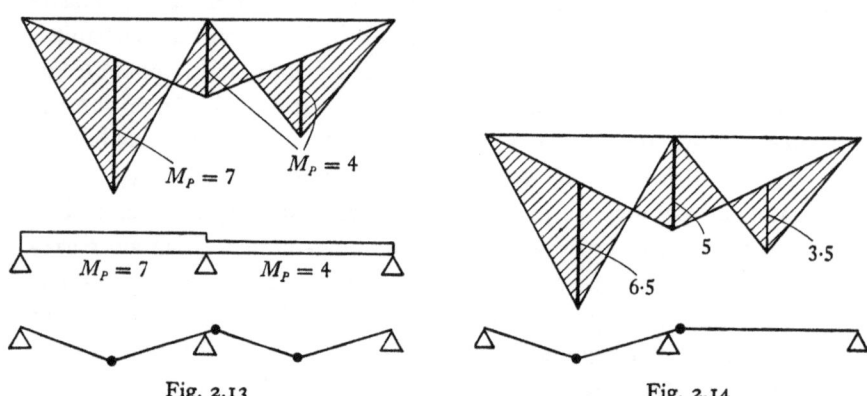

Fig. 2.13   Fig. 2.14

For example, two beams having $M_p = 6 \cdot 5$ and $M_p = 5$ respectively could be used to carry the loads; the bending-moment diagram is shown in fig. 2.14. If it be assumed that the weight of a member is proportional to its full plastic moment, then the designs of figs. 2.11, 2.13 and 2.14 can be compared as in table 2.1:

Table 2.1

Design	Fig. no.	M_p (left)	M_p (right)	Weight (arbitrary scale)
1	2.11	6	6	12
2	2.13	7	4	11
3	2.14	6·5	5	11·5

On this basis it would seem that the design of fig. 2.13 would give the least weight; the uniform section might, however, provide the cheapest design because of the greater cost of joining spans of different sections. The whole question of design for minimum weight is of great interest, but outside the present scope; it will be treated in vol. 2 of this book.

The opportunity for direct design by plastic methods can be illustrated further by the bending-moment diagram of fig. 2.12, redrawn in fig. 2.15. Had the decision been taken to use a basic section for both spans having $M_p = 4$, then it will be seen that the full plastic moment is exceeded over a length of only 4·21 near the centre of the left-hand span. The section could be reinforced, for example by cover plates designed in accordance with the principles of chapter 1, over this relatively short length of the beam; the reinforcement would have to provide an additional full plastic moment of 3 units. The weight of such a reinforced design would be 9·05, to the same arbitrary scale of table 2.1 above.

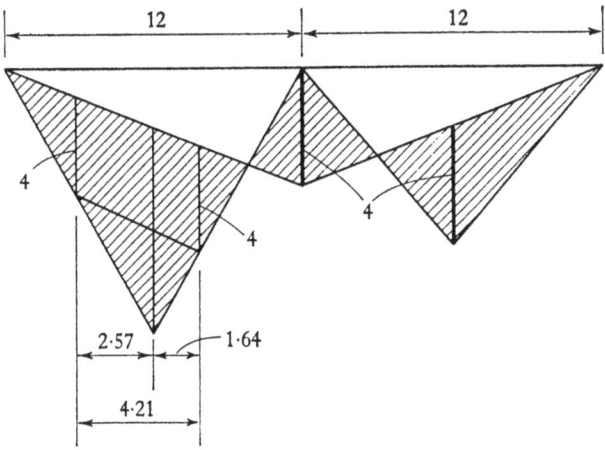

Fig. 2.15

The simple problem of fig. 2.11 has been reinterpreted as one of analysis, rather than design, in fig. 2.16. The beam section is uniform throughout both spans with $M_p = 10$, and the load factor λ is required by which the loads must be multiplied so as just to produce collapse. The two possible mechanisms lead to the two bending-moment diagrams sketched in fig. 2.16; from the geometry of the first diagram, $9\lambda = 15$, or $\lambda = 1·67$, and from the second, $6\lambda = 15$, or $\lambda = 2·5$. The unsafe theorem then indicates that the lower value, $\lambda = 1·67$, is correct.

As another example of collapse *analysis*, the three-span beam of fig. 2.17 will be investigated. Three different uniform cross-sections, having full plastic moments of 12, 14 and 16 respectively, are used for the three spans, and a load of 4λ is applied to the centre of length of each span. The value of the factor λ is required that will bring about collapse. The bending-moment diagrams corresponding to collapse in each span are sketched in the figure; it should be noted that when a

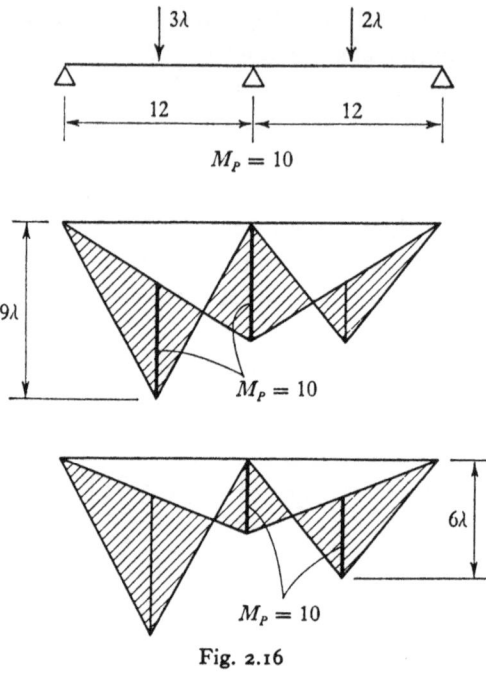

Fig. 2.16

plastic hinge forms at a junction between two members, it must of course form in the weaker member. Thus if collapse occurs in the left-hand span, the moment under the load and at the first internal support will both have the full plastic value 12; from the geometry of the diagram, $9\lambda = 18$, or $\lambda = 2 \cdot 00$. There is insufficient information to complete the bending-moment diagram; the beam system has two redundancies, and only two hinges are formed, leaving the whole system once indeterminate.

Collapse in the centre span involves the formation of three hinges, and the complete bending-moment diagram may be drawn as shown. From the values marked in the diagram, it will be seen that $14\lambda = 27$, or $\lambda = 1 \cdot 93$. For collapse in the right-hand span, $12\lambda = 23$, or $\lambda = 1 \cdot 92$. Of the three possibilities, therefore, collapse in the right-hand span leads to the lowest value of λ, and hence gives the correct mechanism.

Since all three possible mechanisms have been investigated, it is certain that the correct mechanism has been found. If the loads are imagined to be slowly increased, that is, the value of λ slowly increased from the working value of unity, the right-hand span will collapse when λ reaches the value $1 \cdot 92$, before the central or the left-hand span.

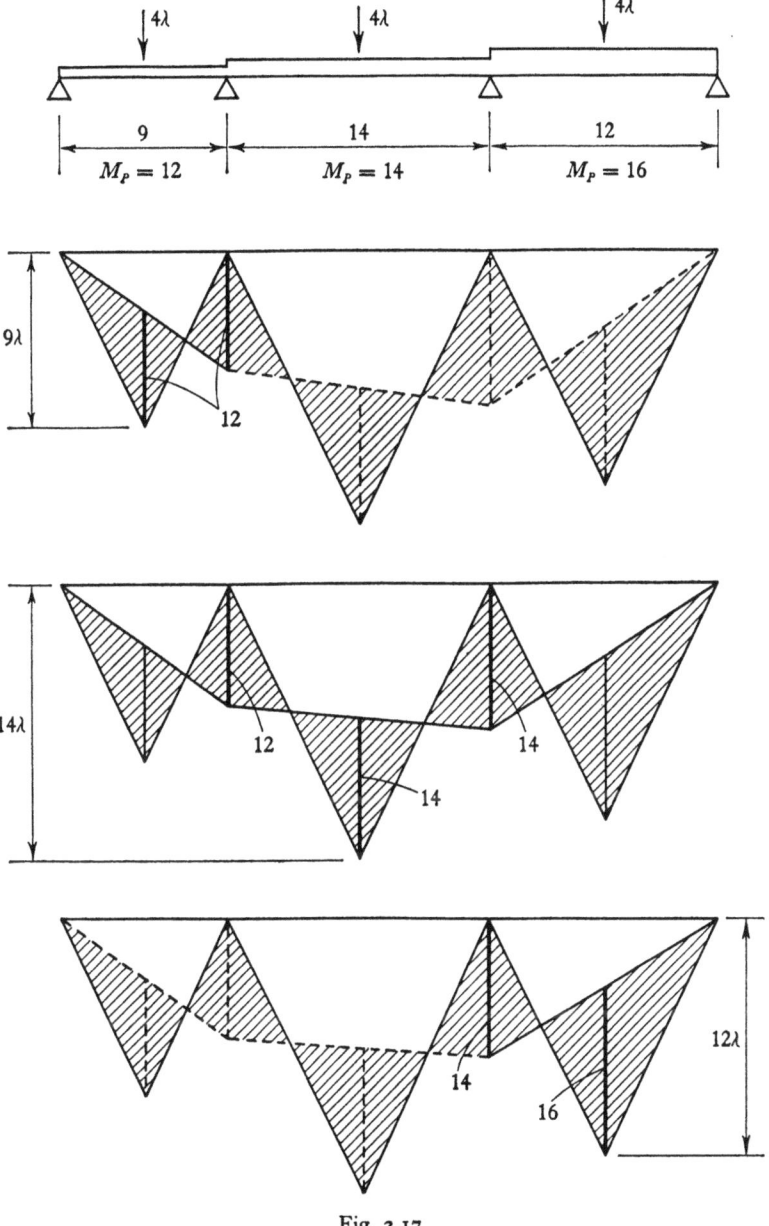

Fig. 2.17

However, the *complete* collapse bending-moment diagram cannot be constructed on the basis of this analysis, and it cannot be checked at this stage that the yield condition is satisfied throughout the beam; one redundancy remains undetermined. This matter will be discussed further in chapter 3.

A very similar point arises in the consideration of a uniform continuous beam, like a roof purlin for example, carrying a uniformly distributed load over several equal spans. It seems clear that such a beam system, fig. 2.18,

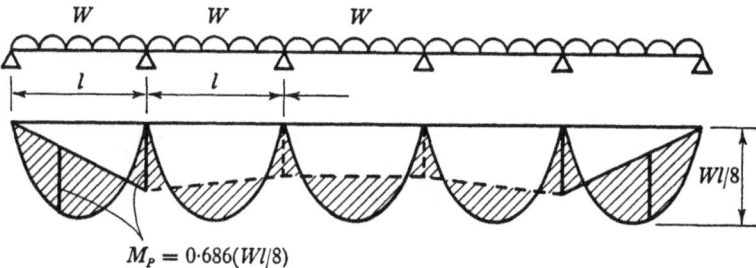

$$M_P = 0.686(Wl/8)$$

Fig. 2.18

will collapse in the end spans only, for which $M_p = 0.686(Wl/8)$ from equation (2.8). The internal spans remain statically indeterminate, although it may be seen by eye that it is, at least, possible to draw in a reactant line such that all bending moments are less than M_p (i.e. so that the yield condition is satisfied). It turns out that a demonstration of the possibility of satisfying the yield condition is all that is necessary to establish the correctness of such an analysis; again, this point will be discussed in chapter 3.

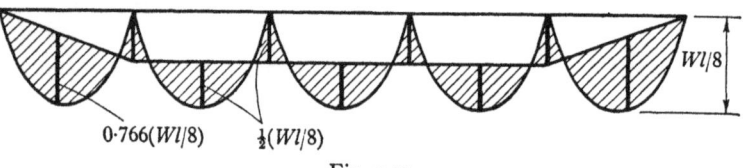

$$0.766(Wl/8) \qquad \tfrac{1}{2}(Wl/8)$$

Fig. 2.19

Figure 2.19 shows an alternative design for the continuous beam, in which the central spans have been given the minimum possible section $M_p = \tfrac{1}{2}(Wl/8)$; this implies a slight increase in the full plastic moment of the end spans. If the weight of the beam is again taken as proportional to the value of M_p, then the following table may be drawn up:

Table 2.2

Design	Fig. no.	Weight (arbitrary scale) for			
		n spans	$n = 3$	$n = 4$	$n = 5$
1	2.18	$0.686n$	2·06	2·74	3·43
2	2.19	$0.5n + 0.532$	2·03	2·53	3·03

As before, the lighter design may not be the cheaper.

2.3 Travelling loads

The concept of the influence line does not carry over to plastic analysis. It is possible to analyse a structure under the action of a single point load placed in an arbitrary position, but this information cannot then be used for the solution of the problem of a train of loads acting on the structure. This is because the plastic solution involves non-linear behaviour, and results cannot be added simply as they can for the corresponding elastic problem; in other words, the principle of superposition for linear elastic structures no longer holds in the plastic range.

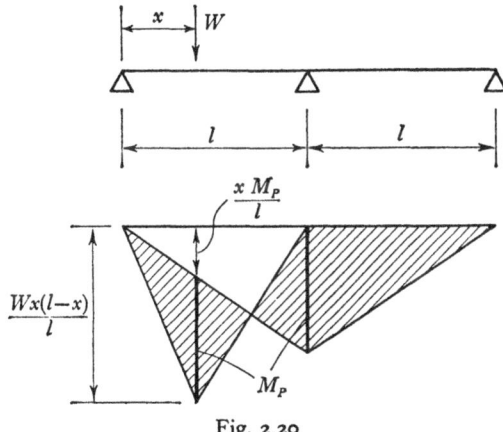

Fig. 2.20

As an example, the two-span beam of uniform section shown in fig. 2.20 will be analysed. For the point load W placed in a general position distant x from the end support, examination of the collapse bending-moment diagrams furnishes the relation

$$M_p + \frac{x}{l} M_p = \frac{Wx(l-x)}{l},$$

or
$$M_p = \frac{Wx(l-x)}{(l+x)}. \qquad (2.10)$$

By differentiation of this expression, it is found that the largest value of M_p occurs when
$$x(l-x) = (l-2x)(l+x),$$

that is
$$x = (\sqrt{2}-1)\,l = 0.414l. \qquad (2.11)$$

Thus the worst position of the load is slightly off-centre in one span; the corresponding value of M_p is

$$M_p = (3-2\sqrt{2})\,Wl = 0.686(Wl/4). \qquad (2.12)$$

(To guard against *incremental collapse* under repeated passages of a point load, the value of M_p would have to be increased above the value given by equation (2.12). The topic of incremental collapse is of restricted practical importance and is again outside the scope of this volume; it will be discussed in vol. 2.)

In the above analysis the weight of the beam itself and the influence of other loads has been neglected; there is no great difficulty in analysing similar beams under more complex loading systems. For example, the single point load of fig. 2.20 could be replaced by two loads separated by a small distance, representing wheel loads from a crane.

2.4 The effect of shear force

In all the above examples of collapse of beams it has been assumed that the value of the full plastic moment at a hinge is unaffected by the shear force acting there. An example of a two-span beam will illustrate how

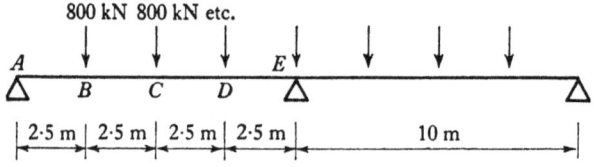

Fig. 2.21

the effect of shear force, already discussed in section 1.6, can be taken into account. The loads shown in fig. 2.21 have been given their working values; it is required to design the beam to collapse at a load factor of 1.75. Due to symmetry, one half only of the beam need be considered, fig. 2.22.

64

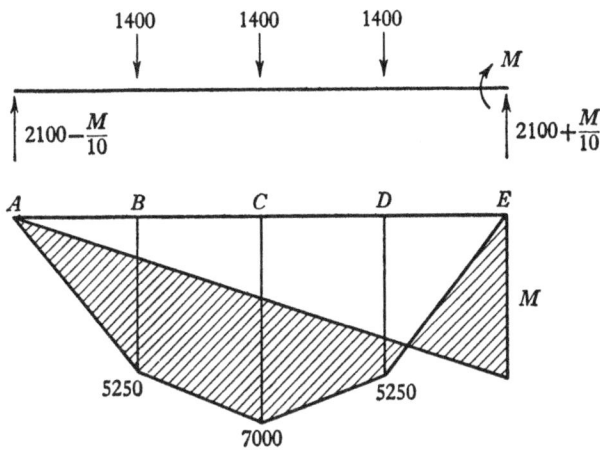

Fig. 2.22

The design of this girder is very open, since by variation of both flange and web areas there is a wide choice of possible bending moments M acting at the central support E. If the beam were uniform, and ignoring the effect of shear force, the value of M would be equal to that of M_p, where $M_p = \frac{2}{3}(7000) = 4667$ kNm. Suppose, as a first trial, that it is assumed that the value of M will be reduced to 4000 kNm due to the effect of shear force. Then the bending-moment and shear-force diagrams for design of the girder will be as shown in fig. 2.23.

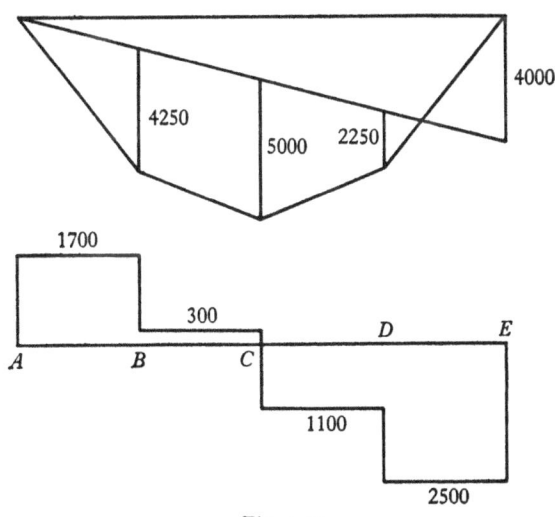

Fig. 2.23

65

The shear force at section E has a value 2500 kN; from section 1.7 it was seen that the shear stress on the web of an I-girder cannot exceed 144 N/mm² if the yield stress in tension is 250 N/mm². Thus a minimum web area of $2500000/144 = 17360$ mm² must be provided. If the web is unstiffened, its depth to breadth ratio must not exceed 85, so that a web of dimensions 1200×15 mm will be tried, having an area of 18000 mm². In the absence of shear force, the web will have a full plastic moment

$$M_{w_0} = \tfrac{1}{4}(15)(1200)^2(250) \times 10^{-6} = 1350 \text{ kNm.} \qquad (2.13)$$

The actual shearing stress at E is $2500000/18000 = 139$ N/mm², so that, from the work of section 1.7,

$$M_w = 1350 \sqrt{\{1 - 3(\tfrac{139}{250})^2\}} = 360 \text{ kNm.} \qquad (2.14)$$

Thus at E the flanges are required to contribute an amount

$$4000 - 360 = 3640 \text{ kNm}$$

to the value of the full plastic moment.

Assuming flanges each of area A and thickness 30 mm, then equation (1.39) gives, with $\sigma_0 = 250$ N/mm²,

$$3640000 = (1 \cdot 230)(A)(250) \qquad (2.15)$$

or $$A = 11820 \text{ mm}^2,$$

say 410 mm $\times$ 30 mm, for which

$$M_f = (1 \cdot 230)(12300)(250) \times 10^{-3} = 3780 \text{ kNm.} \qquad (2.16)$$

Equations (2.14) and (2.16) give a total M_p of 4140 kNm compared with the value of 4000 required in fig. 2.23.

The same web and flange may be used throughout the girder; at the central cross-section, C, the shearing stress is

$$1100000/18000 = 61 \text{ N/mm}^2,$$

so that $$M_w = 1350 \sqrt{\{1 - 3(\tfrac{61}{250})^2\}} = 1220 \text{ kNm.} \qquad (2.17)$$

Equations (2.16) and (2.17) show that the section can develop a full plastic moment at C of value 5000 kNm, agreeing exactly with the value required by fig. 2.23.

Before going on to modify this design, it should be noted that the girder with 1200×15 mm web and 410×30 mm flanges will be *safe* under the factored loads of fig. 2.22. Figure 2.23, although drawn for an arbitrarily assumed value of the bending moment M (4000 kNm), gives a proper equilibrium distribution of bending moments and shear forces for the beam. Flange and web areas have been provided such that this equilibrium state satisfies the yield condition, that is, the full plastic moment of every section of the girder, allowing for shear force, is greater than the corresponding moment at the same section given in fig. 2.23. Thus this particular girder cannot possibly collapse under the loads (already factored) of fig. 2.22; put another way, the actual load factor against collapse will be marginally higher than 1·75.

There was a small margin at cross-section E, 4140 against 4000 kNm, and there is scope for some slight theoretical improvement in the design by reducing the flange area. If the full plastic moments of the web at E and C are assumed to remain unchanged at the values of equations (2.14) and (2.17), and if M_f denotes the new plastic moment of the flanges, then examination of fig. 2.22 shows that:

$$\text{at } E \qquad\qquad M = M_f + 360,$$
$$\text{and at } C, \qquad 7000 - \tfrac{1}{2}M = M_f + 1220. \qquad (2.18)$$

These two equations solve to give $M = 4090$, $M_f = 3730$ kNm. The corresponding values of shear force marked in fig. 2.23 will now be in error by 9 kN each, which is negligible. The full plastic moment of 3730 kNm can just be provided by 405×30 mm flanges, so that a uniform girder having these flanges and a 1200×15 mm web will just carry the design loads of fig. 2.21 at a load factor of 1·75.

The girder is long enough in this example for it to be economical to vary the cross-section; one such design could result from the bending-moment diagram of fig. 2.24. Here the reactant line has been positioned to give a high bending moment at E and equal moments at B and C. It must be emphasized that, in a design of this type, the positioning of the reactant line is at the choice of the designer, and he can quickly try several alternatives to find which is most economical.

Working from fig. 2.24, if a web $1200 \times 17·5$ mm is used, then the mean shear stress at E is $2\,800\,000/21\,000 = 133$ N/mm². The web at E can thus carry a bending-moment 610 kNm, and the flanges are required to contribute a bending moment of 6390 kNm. Flanges of section 410×50 mm and of yield stress 250 N/mm² have (from equation (1.39)),

$$M_f = (1·250)(20\,500)(250) \times 10^{-3} = 6410 \text{ kNm.} \qquad (2.19)$$

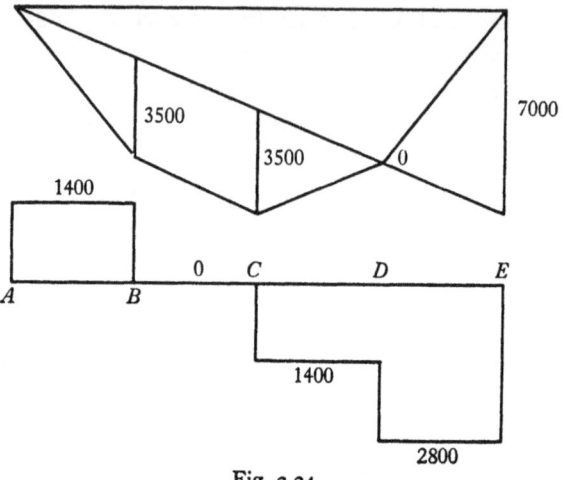

Fig. 2.24

At sections B and C the shear force is 1400 kN, and the corresponding full plastic moment of the web is 1390 kNm, leaving a moment of 2110 kNm to be provided by the flanges; 280 × 25 mm flanges have

$$M_f = (1\cdot225)(7000)(250)(10^{-3}) = 2140 \text{ kNm.} \qquad (2.20)$$

Thus the non-uniform girder of fig. 2.25 would also carry the design loads, and should show some economy in cost as well as in weight compared with the girder of uniform section.

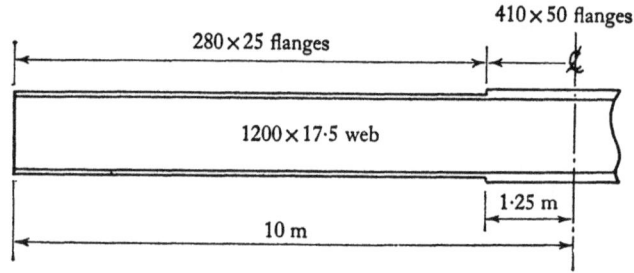

Fig. 2.25

2.5 The work equation

The sketching of bending-moment diagrams, composed of free and reactant parts, is an almost essential preliminary to the analysis or design of even quite a complex frame. In many cases, as has been seen, it is possible to make plastic calculations directly from such sketches, and the designer is afforded an insight into structural behaviour that he would be denied if he used a purely analytical technique without the aid of graphics. However, even if the bending-moment diagrams are simple to sketch, the final calculations are often most conveniently made analytically. The fundamental tool here is the equation of virtual work, of which a fuller account will be given in chapter 3.

For the moment, it is convenient to regard the virtual work equation in the greatly restricted sense of a simple energy balance for the structure in the collapse state. Suppose a frame is on the point of collapse by the formation of plastic hinges, so that a small deformation of the collapse mechanism can occur at constant values of the applied loads. During such a small deformation, a typical load W will do work by moving through a certain distance, say δ; the magnitude of the work done is simply $\Sigma W\delta$, where the summation extends over all the loads on the frame.

This work done by the external loads will be absorbed in the rotating plastic hinges, which turn through certain angles, say θ, at constant bending moments M_p. The work dissipated in the hinges is therefore simply $\Sigma M_p\theta$, so that

$$\Sigma W\delta = \Sigma M_p\theta. \tag{2.21}$$

As a trivial example, the collapse of a simply-supported beam carrying a central point load, fig. 2.26, may be investigated. If the elastic deformations of the beam are taken to be so small that the beam may be assumed to be straight, except at the 'kink' caused by the plastic hinge, then it will be seen that if each half of the beam has a rotation θ, the hinge rotation at the centre is 2θ, and the deflexion under the load is $\frac{1}{2}l\theta$. Equation (2.21) then gives

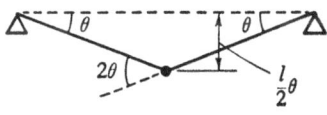

Fig. 2.26

$$W(\tfrac{1}{2}l\theta) = M_p(2\theta),$$

or $$M_p = Wl/4. \tag{2.22}$$

It may be noted that the deflexion δ of the loading point was expressed directly in terms of θ, so that θ could be cancelled on both sides of the work equation. This will always be true for a mechanism of one degree of freedom, that is, for a mechanism which can be completely specified in terms of one arbitrary parameter.

The apparent restriction implied by ignoring the elastic compared with the plastic deflexions disappears if the equation of *virtual* work is used; this is discussed in chapter 3. Alternatively, the rotations θ might be thought of as incremental displacements in the collapse state; in fig. 2.7, for example, deformations could be measured from the point at which three hinges form, so that the elastic components of the deformation do not enter the analysis. However, it is probably best for the time being to regard equation (2.21) as a 'real' work equation in which elastic terms are so small as to be negligible; all collapse mechanisms will be drawn with straight members between hinge points.

Figure 2.27 shows the collapse mechanism for a fixed-ended beam of uniform-section carrying a central point load. The application of equation (2.21) gives

$$W(\tfrac{1}{2}l\theta) = (M_p)(\theta) + (M_p)(2\theta) + (M_p)(\theta),$$

or
$$M_p = Wl/8. \qquad (2.23)$$

On the right-hand side, the terms for each of the three hinges have been displayed separately; since plastic work dissipated at a hinge is always positive, the numerical values of the hinge rotations could have been simply summed to 4θ, the total work dissipated being $(M_p)(4\theta)$.

A consistent sign convention is needed for the full application of the virtual work equation, and for other computational methods dealt with later in this chapter. Hogging bending moments will be denoted positive, so that, for the fixed-ended beam of fig. 2.27, the plastic hinge moments acting at the ends of the beam will be positive. The sagging central hinge will form under a negative bending moment. These concepts of hogging and sagging require modification when applied to frames rather than

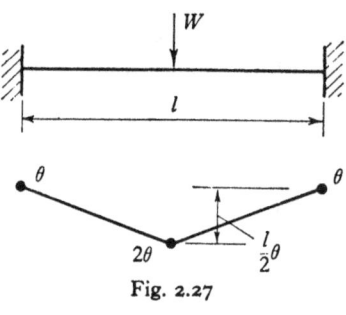

Fig. 2.27

beams, and this will be noted in due course. The essential for present purposes is that, whatever sign convention is used, the signs of hinge rotations must be taken to be the same as the sign of the corresponding full plastic moments. Thus, on the right-hand side of the work equation (2.21), a positive hinge rotation will be multiplied by a positive value of full plastic moment, and a negative rotation by a negative moment.

In either case the product is positive, so that each term on the right-hand side of equation (2.21) is essentially positive.

Thus in fig. 2.28, which shows the collapse mode (identical with that of fig. 2.27) for a fixed-ended beam under uniformly distributed loading, the central hinge rotation has been noted as -2θ. The uniform load W moves through an average distance $\frac{1}{4}l\theta$, so that

$$W(\tfrac{1}{4}l\theta) = M_p(4\theta),$$

or
$$M_p = Wl/16. \tag{2.24}$$

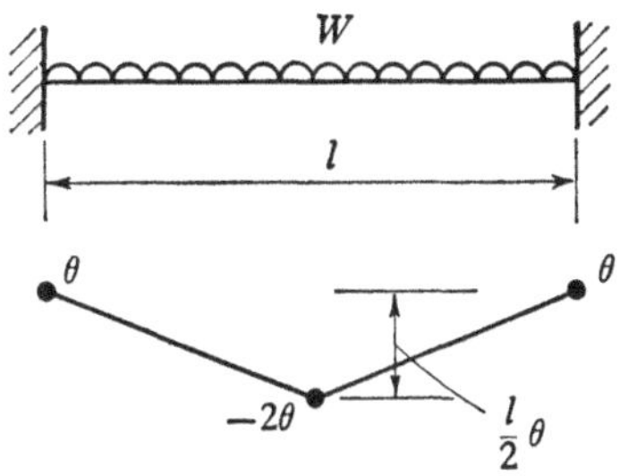

Fig. 2.28 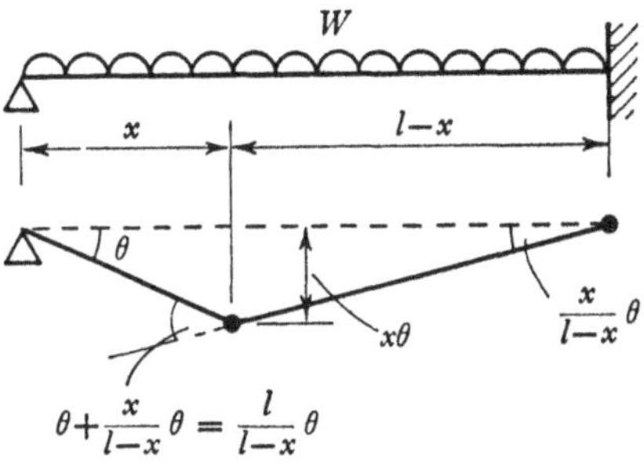   Fig. 2.29

The work equation may be used to solve the problem of the propped cantilever under uniformly distributed load, whose solution was first sketched in fig. 2.10. The difficulty in this problem lay in the exact location of the sagging plastic hinge. In fig. 2.29, this hinge has been placed at an unknown cross-section distant x from the prop, and the collapse mechanism will be as sketched. If θ denotes the rotation of the beam at the prop, then the deflexion at the position of the sagging hinge will be $x\theta$, so that the hinge discontinuity at the right-hand end of the beam must be $x\theta/(l-x)$. Finally, the hinge rotation at the sagging hinge must be the sum of the rotations of the two portions of the beam, leading to the value $l\theta/(l-x)$. Since the uniform load W moves through an average distance of $\frac{1}{2}x\theta$, application of equation (2.21) gives

$$W(\tfrac{1}{2}x\theta) = M_p\left[\frac{l}{l-x}\theta + \frac{x}{l-x}\theta\right],$$

or
$$M_p = \frac{Wx(l-x)}{2(l+x)}. \tag{2.25}$$

The unsafe theorem may now be used to determine the value of x. Any value of x inserted into equation (2.25) will lead to a value of M_p;

that value of x will be correct which gives the greatest value of M_p (i.e. the safest design). Equation (2.25) is of exactly the same form as equation (2.10); the maximum occurs for $x = (\sqrt{2}-1)\,l$, for which

$$M_p = (\tfrac{3}{2}-\sqrt{2})\,Wl = 0\cdot686(Wl/8). \qquad (2.26)$$

The example of fig. 2.17 may be examined again by using the work equation. The three possible (alternative) collapse mechanisms are shown in fig. 2.30; note that, at each support, the hinges are located

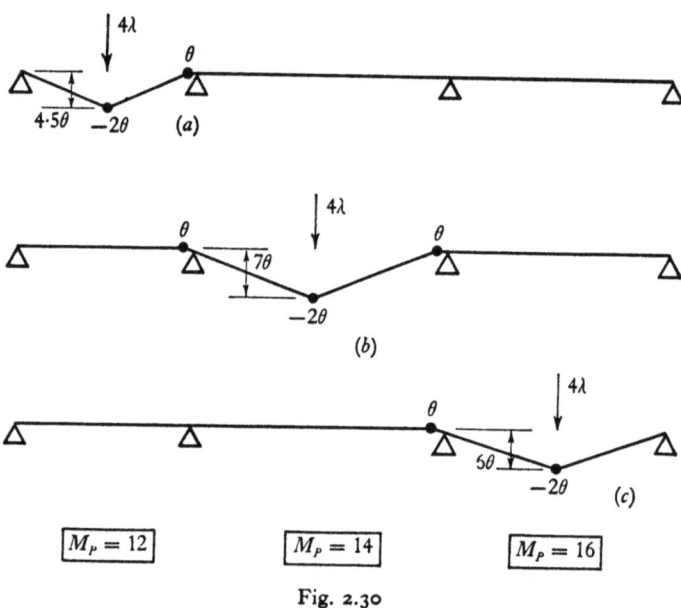

$$M_p = 12 \qquad M_p = 14 \qquad M_p = 16$$

Fig. 2.30

correctly in the weaker member. For each of the three mechanisms, labelled (a), (b) and (c) in fig. 2.30, the collapse equations may be written, from which the load factor λ may be determined:

$$
\begin{aligned}
&(a) \quad 4\lambda(4\cdot5\theta) = (12)(3\theta) &&(\lambda = 2\cdot00);\\
&(b) \quad 4\lambda(7\theta) \;= (12)(\theta)+(14)(3\theta) &&(\lambda = 1\cdot93);\\
&(c) \quad 4\lambda(6\theta) \;= (14)(\theta)+(16)(2\theta) &&(\lambda = 1\cdot92).
\end{aligned}
\qquad (2.27)
$$

These results are, of course, identical with those obtained before, and show that the right-hand span is critical.

Graphical methods will continue to be used where they are appropriate, but it will be found that the work equation saves much labour of calculation. Even when the work equation is used, however, bending moment diagrams should be drawn to help in the analysis and to check the calculations; the diagrams of fig. 2.17 are much more informative than those of fig. 2.30, even if they do lead to exactly the same expressions for the load factor.

2.6 Rectangular portal frames

It has been assumed implicitly in the above discussion of simple beam systems that failure is confined to mechanisms resulting from the formation of plastic hinges. These ideas can be applied unchanged to the problem of design and analysis of frames composed of beams and columns, provided that certain precautions are taken with respect to the columns. The usual assumption for frame analysis will be made that external loads are resisted by bending of the members, and that shear force and axial load cause only secondary effects. Thus while the full plastic moment of a column can be adjusted to allow for axial load, as in section 1.6, it will be assumed that there is no question of an individual column becoming unstable and thus initiating catastrophic collapse of the structure as a whole.

Simple plastic theory indeed cannot be applied directly to structures whose primary failure mode is one of instability. Even the simple beam system must be assumed to be restrained so that lateral instability cannot occur, and the plastic designer must ensure that columns can develop any plastic hinges required in the analysis without becoming unstable. In the whole of this book it is assumed that each member, and the whole structure, is in stable equilibrium right up to the collapse state. Thus columns forming part of a frame will be treated exactly as if they were beams, with their full plastic moments modified if necessary to allow for the effect of axial load.

The simplest possible frame is the rectangular portal of uniform section with pinned feet, fig. 2.31. This frame has one redundancy, and it is to be expected therefore that two hinges will form at collapse. The locations of these two hinges can perhaps best be observed by constructing the bending-moment diagram. The two loads shown in fig. 2.31 are highly idealized representations of dead plus superimposed roof load and of wind load, but they lead to bending-moment diagrams which are characteristic of more realistic loading, and they serve to give the essential plastic behaviour of the rectangular frame.

In fig. 2.32(a) the right-hand column foot has been allowed to move horizontally, so that the frame is statically determinate and free bending

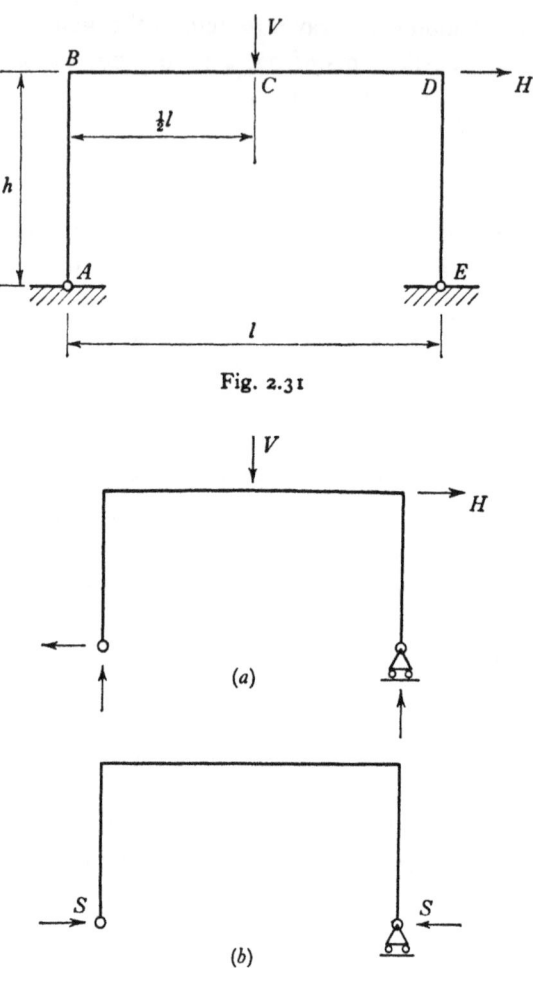

Fig. 2.31

(a)

(b)

Fig. 2.32

moments can be drawn. The single redundancy S of fig. 2.32(b) leads to the reactant bending moments in the frame. The two bending-moment diagrams, free and reactant, corresponding to fig. 2.32, are sketched in fig. 2.33. The frame has been 'opened out' on to a horizontal base line for the purpose of plotting bending moments. Hogging bending moments for the beam are denoted positive, as usual, and plotted above the base line, and the same sign convention is carried into the columns, so that a bending moment producing compression on the inside of the frame is positive.

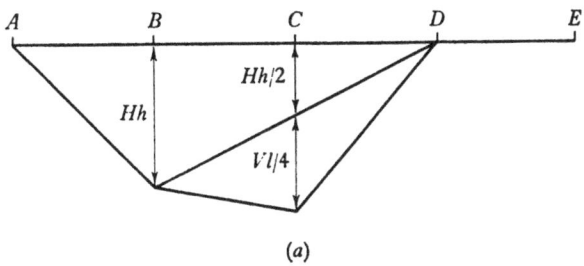

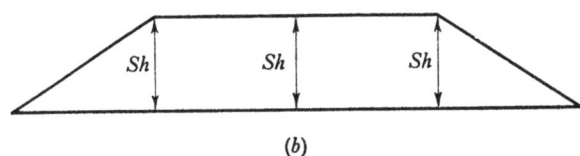

Fig. 2.33

The two diagrams of fig. 2.33 must be superimposed in such a way that a collapse mechanism is formed with two plastic hinges. There are two possibilities, depending on the relative magnitudes of the free bending moments at the sections B and C, and these are sketched in fig. 2.34;

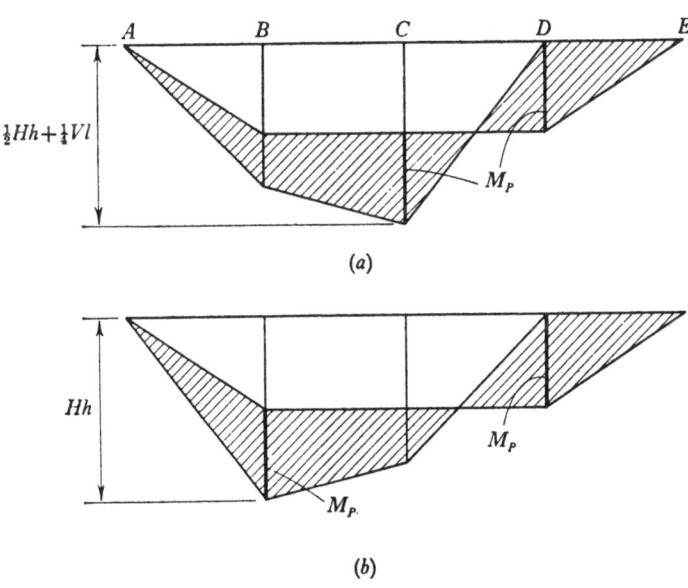

Fig. 2.34

75

it is assumed that the axial load in the columns does not affect significantly the value of the full plastic moment of the uniform section. Working directly from fig. 2.34(a), it will be seen that

$$M_p = (Hh/4)+(Vl/8), \qquad (2.28)$$

while, from fig. 2.34(b), $\qquad M_p = Hh/2.$ $\qquad\qquad (2.29)$

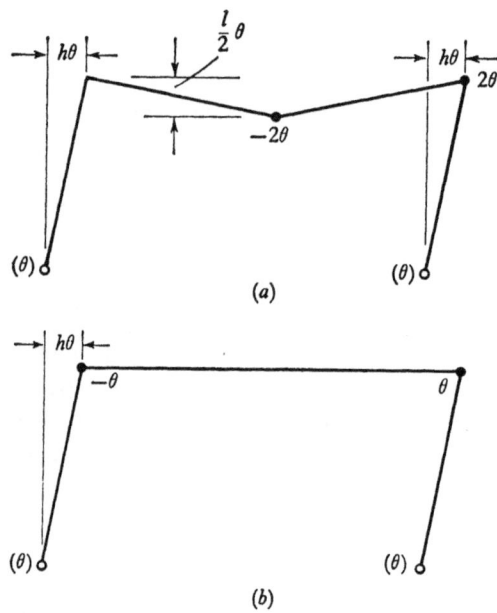

(a)

(b)

Fig. 2.35

The two collapse mechanisms corresponding to the bending-moment distributions of fig. 2.34 are shown in fig. 2.35. The results of equations (2.28) and (2.29) could have been obtained by writing work equations for the collapse mechanisms rather than from an examination of the bending-moment diagrams. For example, fig. 2.35(a) leads to the equation

$$H(h\theta)+V(\tfrac{1}{2}l\theta) = M_p(4\theta), \qquad (2.30)$$

which is identical with equation (2.28).

There are further alternative techniques for deriving collapse relationships such as equation (2.28). Some methods are better than others for the solution of any given problem, but there usually remains an element of personal taste as to which particular technique is used. However, the virtual work equation is clearly best for some types of calculation, while a direct statical approach is best for others; and it is of interest to explore further the use of statics in relation to the pin-based frame.

Suppose, for example, that the collapse mode of fig. 2.35(a) has been guessed to be correct for certain values of the loads and dimensions. The whole frame is statically determinate at collapse, so that an examination of free body diagrams for the various portions of the frame should furnish the collapse equation. Figure 2.36 shows the columns and the

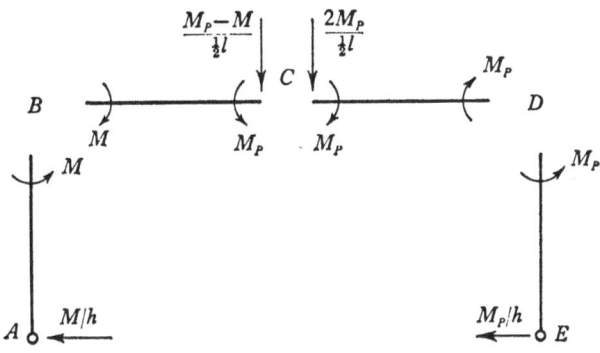

Fig. 2.36

two half-beams displayed separately; the bending moments and forces shown act on the ends of the members. At C and D the full plastic moments have been drawn acting in their proper senses (sagging at C and hogging at D). The bending moment at the corner B is at present unknown, and has been denoted by M. Certain shear forces must act on each of the members of the frame in order to preserve equilibrium, and some of these are shown in fig. 2.36. For column DE, for example, moments about D leads immediately to the value M_p/h for the force acting at the foot of the column.

By consideration of the equilibrium of the whole frame, it is seen that the sum of the shear forces at the column feet must equal the side load H. Thus, from fig. 2.36,

$$Hh = M + M_p. \tag{2.31}$$

Similarly, the two shear forces acting at C, the midpoint of the beam, must sum to the vertical load V acting at that point, so that

$$\tfrac{1}{2}Vl = -M + 3M_p. \tag{2.32}$$

77

Equations (2.31) and (2.32) may be added to eliminate the unknown bending moment M, and the collapse equation (2.28) results.

The bending-moment distribution of fig. 2.36 corresponds to a proper mechanism of collapse, fig. 2.35(a), and the writing of equations (2.31) and (2.32) ensures that equilibrium is satisfied. If the whole solution is to be correct, it is necessary to satisfy also the yield condition, which in this example requires only that the bending moment M at the corner B of the frame must be less numerically than M_p. From equation (2.31),

$$M = Hh - M_p, \tag{2.33}$$

so that the condition $-M_p \leqslant M \leqslant M_p$ implies

$$-M_p \leqslant Hh - M_p \leqslant M_p,$$

or
$$0 \leqslant Hh \leqslant 2M_p. \tag{2.34}$$

Using the value $M_p = Hh/4 + Vl/8$ from equation (2.28), the continued inequality (2.34) becomes

$$Vl \geqslant 2Hh \geqslant 0. \tag{2.35}$$

For $Vl \leqslant 2Hh$ the pure sidesway mode of collapse of fig. 2.35(b) will occur; the condition for the formation of one mode rather than the other can be confirmed by an examination of the bending-moment diagrams of fig. 2.34.

The values of the bending moments at each of the sections at which a hinge might form can be written in general terms from the free and reactant values of fig. 2.33. Since the bending moment at each section is simply the sum of the free and reactant values at that section, then the total moments at sections B, C and D are

$$\left.\begin{aligned} M_B &= -Hh &&+ Sh, \\ M_C &= -\left(\frac{Hh}{2} + \frac{Vl}{4}\right) + Sh, \\ M_D &= &&Sh. \end{aligned}\right\} \tag{2.36}$$

The established sign convention has been followed, in which sagging moments are negative.

On writing the bending moments in this way, it becomes clearer why a regular collapse mechanism involves hinges to the number of one more than the number of redundancies. If collapse occurs by the mode of fig. 2.35(a), for example, with hinges at C and D, then the last two of equations (2.36) may be written

$$\left.\begin{array}{rl} C: & -M_p = -\left(\dfrac{Hh}{2}+\dfrac{Vl}{4}\right)+Sh, \\ D: & M_p = \phantom{-\left(\dfrac{Hh}{2}+\dfrac{Vl}{4}\right)+} Sh. \end{array}\right\} \tag{2.37}$$

The signs of the bending moments at the plastic hinges in equations (2.37) agree with the signs of the hinge rotations in fig. 2.35(a). The unwanted redundancy S may be eliminated from equations (2.37), leaving the single collapse equation (which is, of course, equation (2.28) yet again). In general, if there are a number R redundancies, then $(R+1)$ equations of the type (2.37) may be written, one for each hinge, so that all the redundant quantities may be eliminated leaving a single collapse equation.

The analytical technique illustrated by equations (2.37) is of some importance in the design of relatively simple structures acted upon by real loads, which may themselves be complex. An example is given in section 2.7 below for the fixed-base frame, and again in section 2.8 for the pitched-roof frame.

The two possible collapse mechanisms of fig. 2.35 for the uniform pin-based rectangular frame under idealized loading may be shown graphically on a single diagram. The two collapse equations (2.28) and (2.29) may be plotted as shown in fig. 2.37, in which the two axes correspond to the two loads acting on the frame. This interaction diagram is another example of a yield surface, of which examples were given in sections 1.6 and 1.8 for the formation of plastic hinges under combined loading systems.

Figure 2.37 represents a 'yield surface' for the entire structure, in this case of a frame with given dimensions and known full plastic moment. Any point in the plane of fig. 2.37 represents a certain combination of the external loads V and H. If this point lies on the boundary PQR then collapse is just occurring under the corresponding loads, by the mode of fig. 2.35(a) if the point is on PQ, and by the mode of fig. 2.35(b) if the point is on QR. A point *within* the boundary PQR (on the origin side) represents values of loads for which the structure is safe, i.e. one for which collapse cannot occur. A point outside the boundary represents loads which cannot be carried by the frame.

The point Q in the diagram represents the state at which, in theory, either of the two collapse modes can form, and corresponds to $Vl = 2Hh$ (cf. inequality (2.35)).

Diagrams of the type shown in fig. 2.37 must always be convex round the origin; they can contain no re-entrant angles, for example. Thus, to establish that such a diagram is complete for a more complex frame in which the modes of collapse are not self-evident, and one of which might have been overlooked, it is necessary only to demonstrate that points such as Q are *possible*, in the sense that the yield condition is not violated.

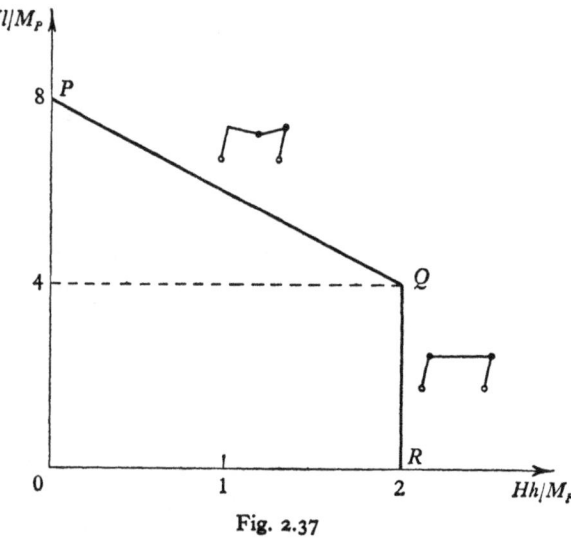

Fig. 2.37

The normality rule is also obeyed by interaction diagrams such as fig. 2.37. If deformation axes (δ_H, δ_V) are imagined to be superimposed on the force axes (H, V), then the normal to the yield surface gives the relative motions of the loading points. For example, the normal to QR in fig. 2.37 is parallel to the H-axis, indicating that the deflexion δ_V of the vertical load V is zero. The normal to PQ gives the ratio $\delta_V/\delta_H = l/2h$, which corresponds with the deflexions noted in fig. 2.35(a).

In the whole of the discussion so far of the pin-based frame it has been assumed that the beams and columns had the same section. Once the free bending-moment diagram has been drawn, however, there is no difficulty in examining the possibilities of a non-uniform design; alternatively, the work equation can again be used. In fig. 2.38(a), for example, in which the side load has been taken as relatively small, a collapse bending-moment diagram has been drawn on the assumption that the column section of full plastic moment M_{p_2} is less than the beam section of full plastic moment M_{p_1}. From the geometry of the diagram, or by writing the work equation for the mechanism of fig. 2.38(b), it will be seen that

$$M_{p_1} + M_{p_2} = (Hh/2) + (Vl/4).\qquad(2.38)$$

It should be noted that there is a minimum value of $M_{p_2} (= Hh/2)$, but that a range of values of M_{p_1} and M_{p_2} satisfying equation (2.38) will give a design of the frame if the free bending-moment diagram is of the general shape sketched in fig. 2.38(a).

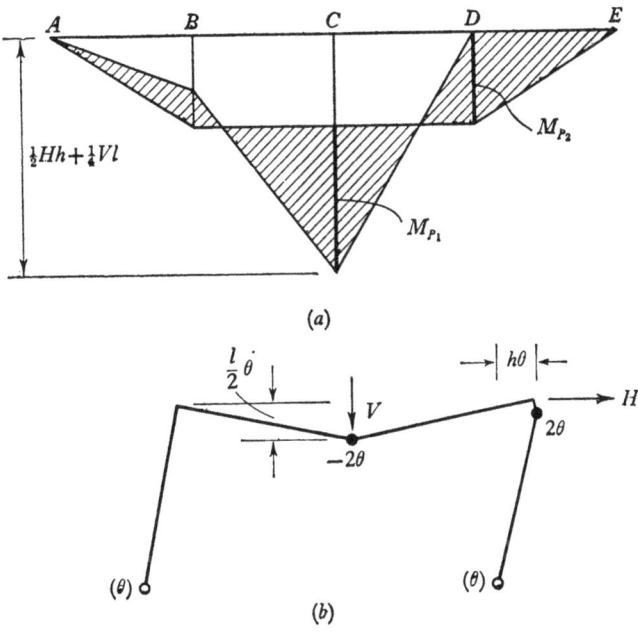

(a)

(b)

Fig. 2.38

As a numerical example, and to illustrate again the ease of direct design by plastic methods, the example of fig. 2.39 will be worked. The two columns have equal section M_{p_2}; the minimum value of M_{p_2} is given, from fig. 2.40(a), by the equation

$$(4\theta) = (M_{p_2})(2\theta),$$

or $\qquad\qquad M_{p_2} = 2.\qquad\qquad(2.39)$

Figure 2.40(a) represents a possible collapse mechanism for a design having $M_{p_2} = 2$, and with M_{p_1} greater than a certain minimum value (this value is in fact 8, as may be confirmed by computing the value of

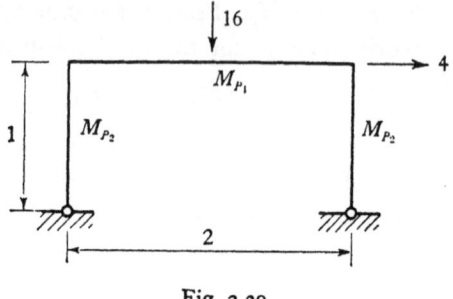

Fig. 2.39

the bending moment at the centre of the beam; however, the minimum value is determined easily by an examination of equation (2.40) below, as will be seen). If the beam section is reduced below this minimum value, then the column section must be increased and the mode of collapse will change to that of fig. 2.40(b). Applying the work equation,

$$2M_{p_1} + 2M_{p_2} = 20. \qquad (2.40)$$

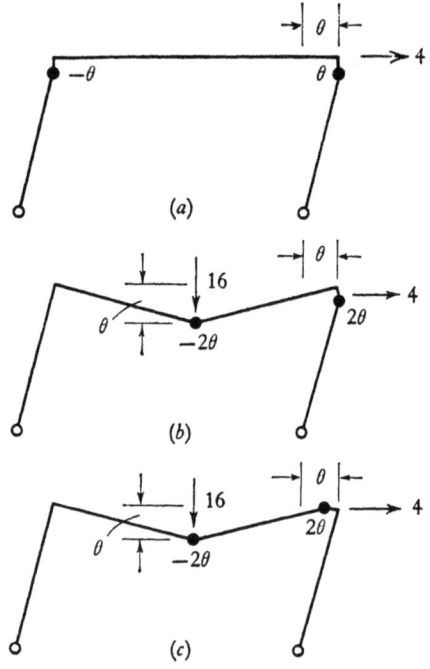

Fig. 2.40

In fig. 2.40(b), it has been assumed that $M_{p_1} > M_{p_2}$, so that the hinge at the right-hand knee forms at the top of the column rather than at the end of the beam. As the beam section continues to be reduced, however, then eventually $M_{p_1} = M_{p_2}$; from equation (2.40), this stage will be reached when the full plastic moment of both sections has the value 5. The collapse mode then changes again to that of fig. 2.40(c), for which

$$4M_{p_1} = 20, \tag{2.41}$$

and this completes the range of possible designs. Other modes of collapse are possible for different values of the loads, for example a sidesway mode with the two plastic hinges formed at the ends of the beam rather than in the columns, but only the three modes of fig. 2.40 are possible for the frame of fig. 2.39.

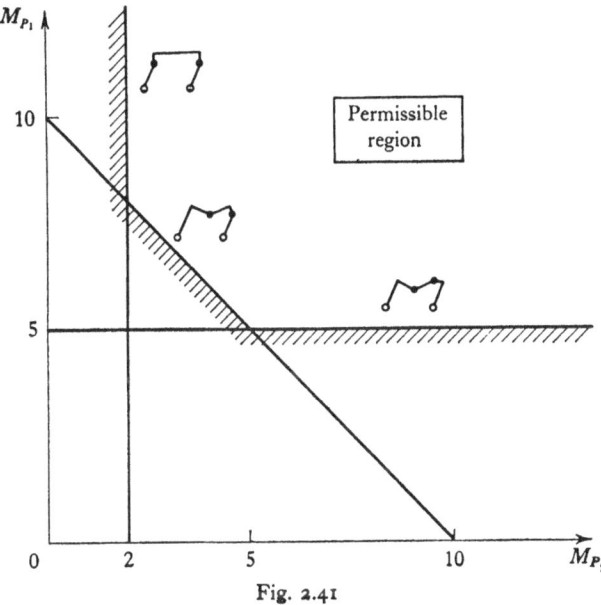

Fig. 2.41

Equations (2.39), (2.40) and (2.41) may be plotted in the *design plane* of fig. 2.41. Any point in this plane represents certain values of M_{p_1} and M_{p_2}, and hence corresponds to a particular design of the frame. If the design point lies on the shaded boundary, then the frame has been designed just to carry the given loads. A point *within* the permissible region represents a safe design of the frame, i.e. one which will not collapse under the given loads. A point on the origin side of the boundary of the permissible region represents a design which will *not* carry the design loads.

The boundary of the permissible region is, in fact, a kind of inverse of the yield surface of the type sketched in fig. 2.37. The yield surface corresponds to a given structure which can be subjected to different combinations of loading, while the boundary of the permissible region in the design plane corresponds to a given loading acting on structure whose design is varied.

In theory, the boundary of the permissible region could be examined to find the most economical frame to carry a given set of loads. In practice, more complex frames give rise to an enormous number of alternative collapse mechanisms, and some general rules must be established to help in the exploration if the labour of calculation is not to be excessive. This will be dealt with in vol. 2.

2.7 The fixed-base rectangular portal frame

The ideas established for the pin-based frame can be carried through to the analysis of the frame with fixed feet. For a frame of uniform section, there are three possible collapse modes under the loading of fig. 2.42.

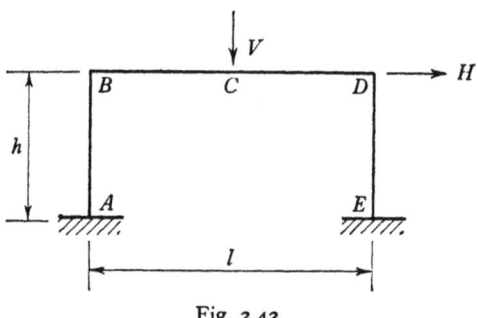

Fig. 2.42

The two modes of the pin-based frame, fig. 2.35, are reproduced in fig. 2.43, with the difference that the hinges at the column feet are plastic hinges and not simple pins. The fixed-base frame has three redundancies, so that four plastic hinges would be expected to form in a regular collapse mechanism, and the modes of fig. 2.43(*a*) and (*b*) obey this rule.

There is, however, a third mode of collapse, fig. 2.43(*c*), which occurs for relatively high values of vertical load and small values of side load. Only three hinges are formed in this incomplete or partial collapse mode, and it is to be expected that, although the frame is collapsing, a degree of statical indeterminacy will remain. This situation has already been encountered with the continuous beam, e.g. in the example of fig. 2.17; the partial collapse mode for the portal frame is analysed more fully below.

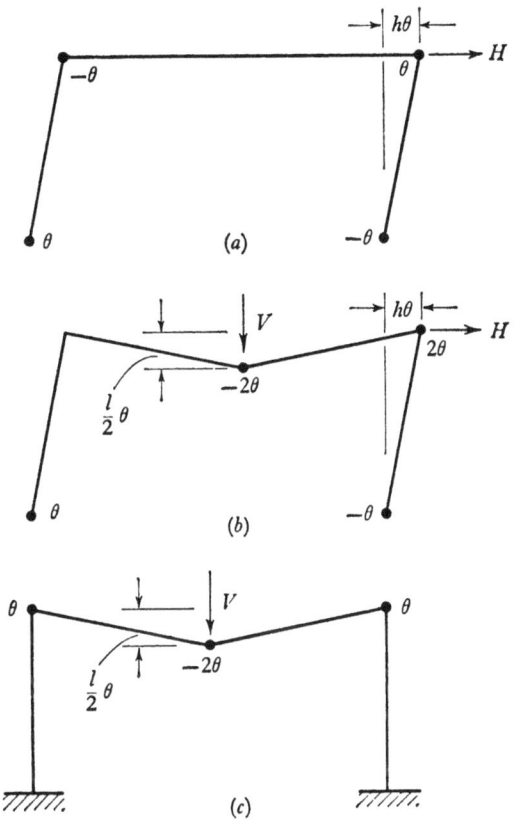

Fig. 2.43

The collapse equations for the three modes can be written by applying the work equation to each of the mechanisms of fig. 2.43:

$$
\begin{aligned}
(a)\quad & Hh && = 4M_p, \\
(b)\quad & Hh + \tfrac{1}{2}Vl && = 6M_p, \\
(c)\quad & \tfrac{1}{2}Vl && = 4M_p.
\end{aligned}
\qquad (2.42)
$$

These three equations can again be displayed on an interaction diagram, fig. 2.44. To confirm that this figure is a complete plot for positive values of V and H, and that there is not some other mode of collapse lying within the convex yield surface, the vertices of the diagram should be examined. It is sufficiently evident that for no load on the beam ($V = 0$) the side load leads to collapse by mode (a) in fig. 2.43 with $Hh = 4M_p$; similarly, for no horizontal load ($H = 0$), $Vl = 8M_p$ is the correct collapse equation.

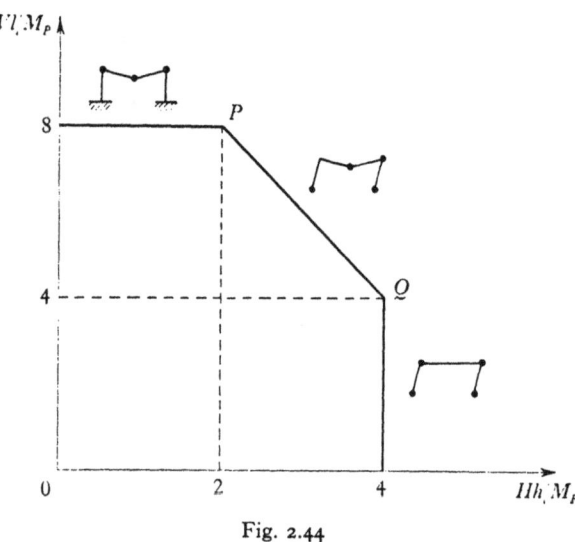

Fig. 2.44

To confirm that points P and Q correspond to states of the frame for which the yield condition is satisfied, it may be noted that the bending-moment diagram consists of straight lines between the hinge positions. Since all five of the possible positions are subjected to the full plastic moment M_p for each of the points P and Q, that moment cannot be exceeded between the five hinge positions. Thus for point Q in fig. 2.44 modes (a) and (b) of fig. 2.43 occur simultaneously. The general bending-moment diagrams for modes (a) and (b) are sketched in fig. 2.45, together

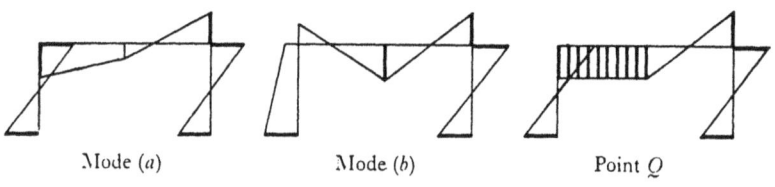

Mode (a)   Mode (b)   Point Q

Fig. 2.45

86

with the unique distribution for the point Q which just satisfies the yield condition (half of the beam is subjected to the full plastic moment).

Thus limits may be assigned to the value of V and H in equations (2.42) for collapse by the various modes:

$$
\begin{aligned}
(a)\quad Hh &= 4M_p \quad (Hh \geqslant Vl), \\
(b)\quad Hh + \tfrac{1}{2}Vl &= 6M_p \quad (Vl \geqslant Hh \geqslant \tfrac{1}{4}Vl \text{ or } Hh \leqslant Vl \leqslant 4Hh), \\
(c)\quad \tfrac{1}{2}Vl &= 4M_p \quad (Vl \geqslant 4Hh).
\end{aligned}
$$

$$(2.43)$$

The limit $Vl \geqslant 4Hh$ for mode (c) to occur rather than mode (b) is apparent from the coordinates of the point P in fig. 2.44; alternatively, the condition may be written from a direct examination of the statics of the collapsing frame. The free-body diagrams for the various portions of the frame have been partly drawn in fig. 2.46; the collapse condition $V = 8M_p/l$ is immediately confirmed by summation of the shear forces acting on the two halves of the beam.

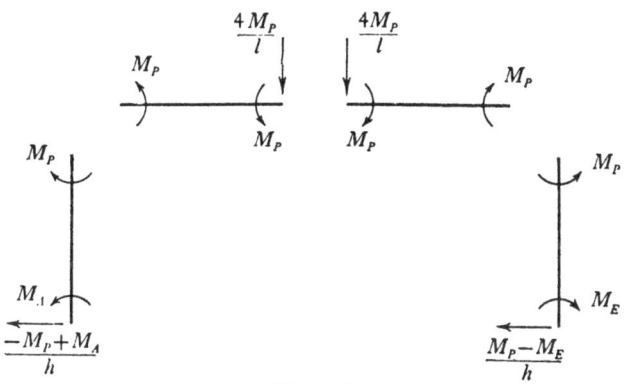

Fig. 2.46

The shear forces at the column feet must sum to the side load H, that is

$$Hh = M_A - M_E. \tag{2.44}$$

Only this one equation can be found from considerations of equilibrium, so that, as suspected, one redundancy is left at collapse (either the value of M_A or that of M_E), despite the fact that a mechanism is formed. If the mode of collapse is correct, however, it is certain that the values of both M_A and M_E must be less, numerically, than M_p, and this fact enables an upper limit to be put on the value of Hh. Thus, from equation

(2.44), it is certain that $Hh \leqslant 2M_p$ for mode (c) to occur, i.e. $4Hh \leqslant Vl$. That mode (c) will *necessarily* occur if the condition $4Hh \leqslant Vl$ is satisfied is a consequence of the safe theorem, which will be stated formally in the next chapter.

The whole of this analysis of the fixed-base frame could have been made by an examination of free and reactant bending-moment diagrams. In fig. 2.47 the frame has been cut through at the centre C of the beam;

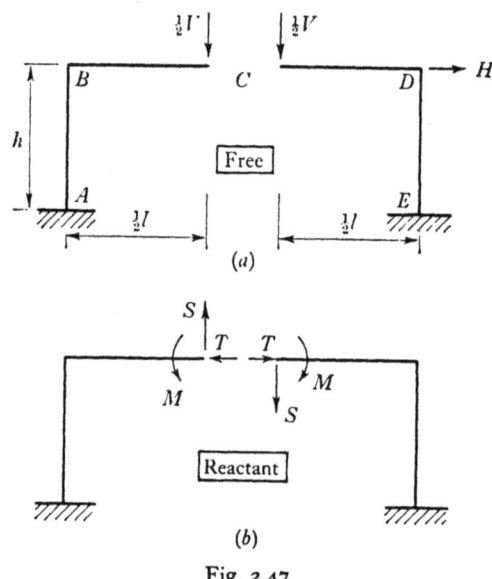

Fig. 2.47

the two bent cantilevers ABC and CDE are statically determinate under any system of loading. By cutting a frame at one cross-section three internal actions are, in general, destroyed; these are a bending-moment M, and also a shear force S and a thrust T. These three unknown quantities are shown acting at the cut in fig. 2.47(b); the fact that a single cut suffices to turn the fixed-base portal frame into a statically determinate structure confirms that it had originally three redundancies, and these three redundancies can quite properly be taken as M, S and T.

In fig. 2.47(a) the vertical load V has been 'cut' into two equal halves in order to preserve some degree of symmetry. This division is otherwise quite arbitrary, since the unknown shear force S acts at the cut. Had all of the load V been placed on one half of the frame, and none on the other, this would merely have affected the final calculated value of S.

Bending-moment diagrams corresponding to fig. 2.47 are shown in fig. 2.48; these two diagrams must be superimposed to give the required distribution corresponding to a possible mode of collapse. The three modes are illustrated in fig. 2.49; in each case it has been assumed that the frame is of uniform section, but it is clear that the reactant line can be positioned for the cases in which beam and column sections differ.

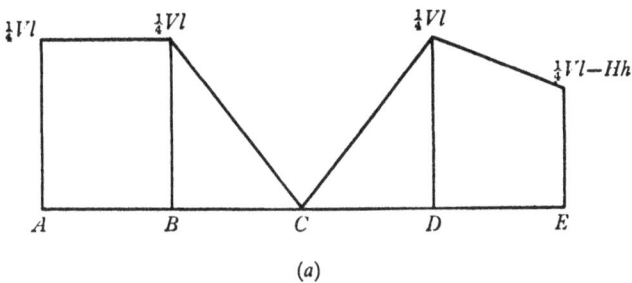

(a)

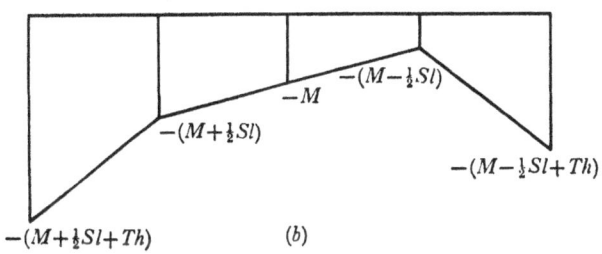

(b)

Fig. 2.48

Exactly as for the pin-based frame, it is often convenient to sketch the diagrams, and then to perform precise calculations by writing the expressions for the bending moments at each of the hinge positions in the form: Free moment + reactant moment $= \pm M_p$. Thus for mode (b), fig. 2.43 and fig. 2.49, in which hinges form at the sections A, C, D and E, the collapse condition may be expressed by the equations

$$\left.\begin{aligned}
A\colon\ & (\tfrac{1}{4}Vl) & -(M+\tfrac{1}{2}Sl+Th) &= M_p, \\
C\colon\ & (0) & -(M) &= -M_p, \\
D\colon\ & (\tfrac{1}{4}Vl) & -(M-\tfrac{1}{2}Sl) &= M_p, \\
E\colon\ & (\tfrac{1}{4}Vl-Hh) - (M-\tfrac{1}{2}Sl+Th) &= -M_p.
\end{aligned}\right\} \qquad (2.45)$$

In these equations, the values of the free and reactant moments have been taken directly from fig. 2.48; as usual, bending moments producing compression on the inside of the frame (hogging moments) are denoted

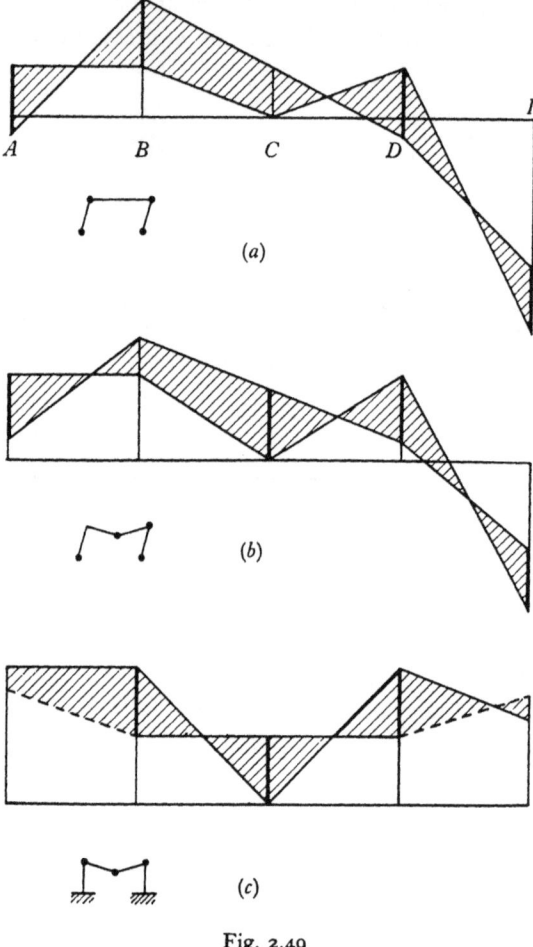

Fig. 2.49

positive. Similarly, the signs of the full plastic moments are consistent with the hinge rotations of fig. 2.43 (b).

The three redundant quantities M, S and T can be eliminated from equations (2.45) to give the collapse equation

$$\tfrac{1}{2}Vl + Hh = 6M_p, \tag{2.46}$$

and substitution back into the equations gives

$$M = M_p = \tfrac{1}{12}Vl + \tfrac{1}{6}Hh,$$
$$\tfrac{1}{2}Sl = -\tfrac{1}{12}Vl + \tfrac{1}{3}Hh,$$
$$Th = \tfrac{1}{6}Vl - \tfrac{2}{3}Hh. \tag{2.47}$$

Using these computed values of the redundancies the bending-moment diagram of fig. 2.49(b) can be constructed completely; on the drawing board it is, of course, necessary only to mark off the computed values of M_p on the free bending-moment diagram, and to complete the resulting reactant diagram as a series of straight lines.

The values of equations (2.47) can be used to check analytically that the bending moment at B satisfies the yield condition. From fig. 2.48,

$$M_B = (\tfrac{1}{4}Vl) - (M + \tfrac{1}{2}Sl) = \tfrac{1}{4}Vl - \tfrac{1}{2}Hh \tag{2.48}$$

on substituting the value of $(M + Sl/2)$ from equations (2.47). The value of M_B must lie between $\pm M_p$, so that

$$\tfrac{1}{12}Vl + \tfrac{1}{6}Hh \geqslant \tfrac{1}{4}Vl - \tfrac{1}{2}Hh \geqslant -\tfrac{1}{12}Vl - \tfrac{1}{6}Hh, \tag{2.49}$$

which can be rearranged to give the condition already found:

$$4Hh \geqslant Vl \geqslant Hh. \tag{2.50}$$

As noted above, the main virtue of writing equations such as (2.45) is that it gives a relatively easy way of computing collapse equations for a real frame acted upon by complex loading systems. The reactant moments for a fixed-base rectangular frame are always given by the expressions of fig. 2.48(b), independently of the loading on the frame; reactant moments are purely functions of the frame geometry. Thus if a table of free bending moments is available for any given loading system, however complex, it is a simple matter to solve a set of equations similar to (2.45).

As a numerical example, the frame of fig. 2.50(a) will be analysed. If the frame is cut as in fig. 2.50(b), the set of free bending moments given in table 2.3 may be calculated. These bending moments are sketched in fig. 2.51.

Suppose first that it is assumed that mode (a) of fig. 2.43 is the correct collapse mode, with hinges formed at sections A, B, D and E. Since the

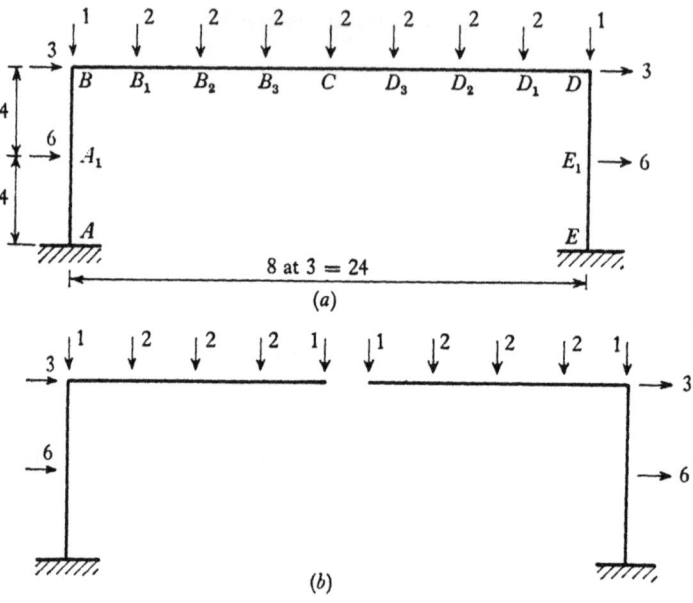

Fig. 2.50

Table 2.3

	A	A_1	B	B_1	B_2	B_3	C	D_3	D_2	D_1	D	E_1	E
Free	96	60	48	27	12	3	o	3	12	27	48	36	o

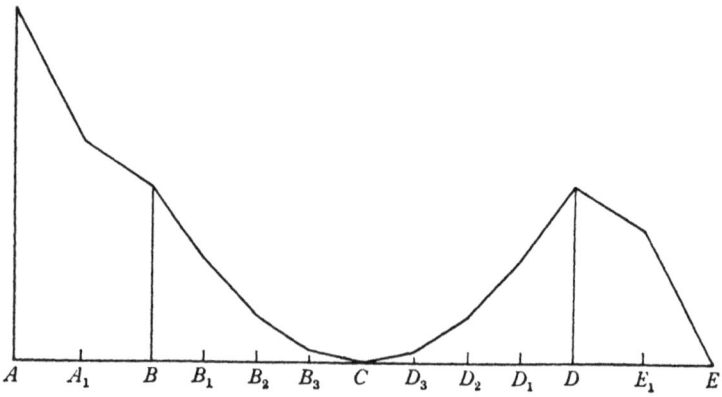

Fig. 2.51

reactant moments are still as given in fig. 2.48(b), where $\frac{1}{2}Sl = 12S$ and $Th = 8T$, the collapse equations are

$$\left.\begin{array}{llll} A: & 96-(M+12S+8T) & = M_p, \\ B: & 48-(M+12S) & = -M_p, \\ D: & 48-(M-12S) & = M_p, \\ E: & 0-(M-12S+8T) & = -M_p. \end{array}\right\} \qquad (2.51)$$

These equations solve to give

$$\left.\begin{array}{rl} M_p & = 24, \\ M & = 48, \\ 12S & = 24, \\ 8T & = 0. \end{array}\right\} \qquad (2.52)$$

Since the reactant diagram consists of straight lines, the full table of moments may be completed as in table 2.4. These values are plotted in fig. 2.52.

Table 2.4

	A	A_1	B	B_1	B_2	B_3	C	D_3	D_2	D_1	D	E_1	E
Free	96	60	48	27	12	3	0	3	12	27	48	36	0
Reactant	−72	−72	−72	−66	−60	−54	−48	−42	−36	−30	−24	−24	−24
Total	24	−12	−24	−39	−48	−51	−48	−39	−24	−3	24	12	−24

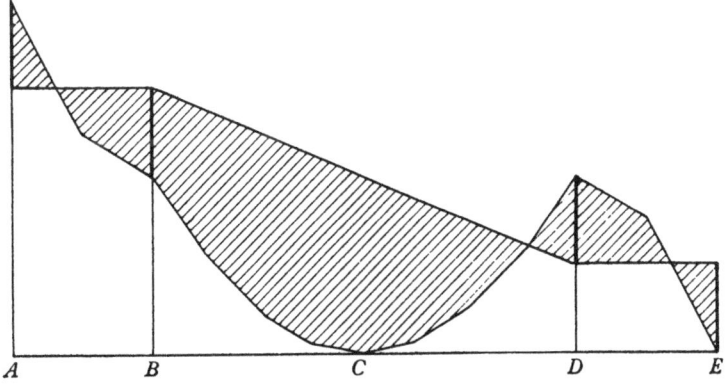

Fig. 2.52

It is clear both from table 2.4 and fig. 2.52 that the guess for the collapse mode was very clumsy, and that even a rough sketch of the bending moments would have indicated the unlikelihood of the collapse mode being one of pure sidesway. From fig. 2.52, it would seem that the collapse mode (b) of fig. 2.43 is much more nearly correct, and this will now be tried as a second guess. The same free bending-moment table (table 2.3) is used, and the reactant moments are again taken from fig. 2.48(b); the new set of equations, replacing (2.51), are:

$$\left.\begin{array}{lll} A: & 96-(M+12S+8T) = M_p, \\ C: & 0-(M) & = -M_p, \\ D: & 48-(M-12S) & = M_p, \\ E: & 0-(M-12S+8T) = -M_p, \end{array}\right\} \quad (2.53)$$

which solve to give

$$\left.\begin{array}{l} M_p = M = 32, \\ 12S = 16, \\ 8T = 16. \end{array}\right\} \quad (2.54)$$

Table 2.5

	A	A_1	B	B_1	B_2	B_3	C	D_3	D_2	D_1	D	E_1	E
Free	96	60	48	27	12	3	0	3	12	27	48	36	0
Reactant	−64	−56	−48	−44	−40	−36	−32	−28	−24	−20	−16	−24	−32
Total	32	4	0	−17	−28	$\boxed{-33}$	−32	−25	−12	7	32	12	−32

The bending moments are now given in table 2.5, and the final line of the table confirms that the calculations have been made correctly, in that plastic hinges are formed at the assumed positions A, C, D and E. However, it will be seen that the collapse mode is not yet quite correct, since the yield condition is violated at the section B_3 just to the left of the centre of length of the beam. It is almost certain that the correct mode is that sketched in fig. 2.53(a), and this will be confirmed by a final calculation:

$$\left.\begin{array}{lll} A: & 96-(M+12S+8T) = M_p, \\ B_3: & 3-(M+ 3S) & = -M_p, \\ D: & 48-(M-12S) & = M_p, \\ E: & 0-(M-12S+8T) = -M_p. \end{array}\right\} \quad (2.55)$$

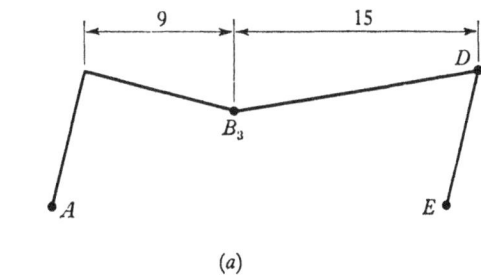

(a)

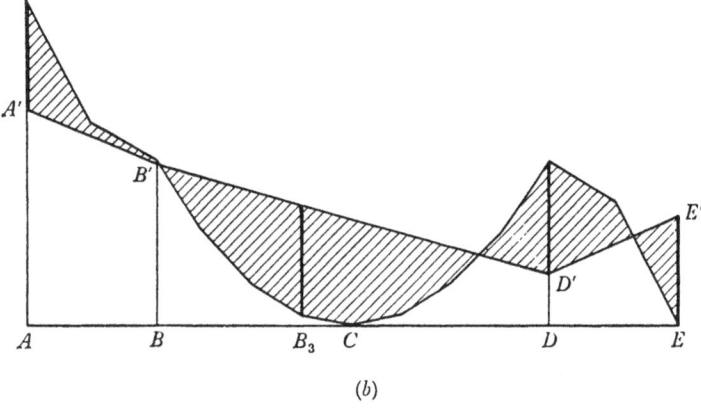

(b)

Fig. 2.53

The value $-(M+3S)$ for the reactant moment at the section B_3 should be apparent from fig. 2.48(b); equations (2.55) solve to give

$$\left.\begin{aligned} M_p &= 32\cdot3, \\ M &= 31\cdot4, \\ 12S &= 15\cdot7, \\ 8T &= 16\cdot6, \end{aligned}\right\} \tag{2.56}$$

and the final table of moments is given in table 2.6. This table confirms that the yield condition is satisfied, and that $M_p = 32\cdot3$ gives the correct value of full plastic moment for a uniform frame.

Even though the first guess, fig. 2.52, was so bad, the corresponding table 2.4 would at least furnish bounds on the value of M_p required for the design of a uniform frame. The value $M_p = 24$ is of course an unsafe estimate; the largest value of bending moment $(M_{B_3} = 51)$ furnishes an upper bound to the value of M_p, so that $51 \geqslant M_p \geqslant 24$. Alternatively, table 2.4 would equally well serve as a basis for the design of a non-

95

uniform frame; the columns could be given their minimum full plastic moments of 24 units, and the beam a section having a full plastic moment of 51 units.

Table 2.6

	A	A_1	B	B_1	B_2	B_3	C
Free	96	60	48	27	12	3	0
Reactant	−63·7	−55·4	−47·1	−43·2	−39·2	−35·3	−31·4
Total	32·3	4·6	0·9	−16·2	−27·2	−32·3	−31·4

	D_3	D_2	D_1	D	E_1	E
Free	3	12	27	48	36	0
Reactant	−27·5	−23·5	−19·6	−15·7	−24·0	−32·3
Total	−24·5	−11·5	7·4	32·3	12·0	−32·3

The particular design example which has just been discussed is probably best solved on the drawing board; if the degree of accuracy is thought to be insufficient (although even a little care will lead to extremely accurate values), then calculation can be used when the correct mode of collapse has been established graphically. Thus, on a plot of the free bending moments, fig. 2.51, must be superimposed by trial and error a proper reactant line of the form of fig. 2.48(*b*), redrawn in fig. 2.54. There are two essential features of the diagram of fig. 2.54; it

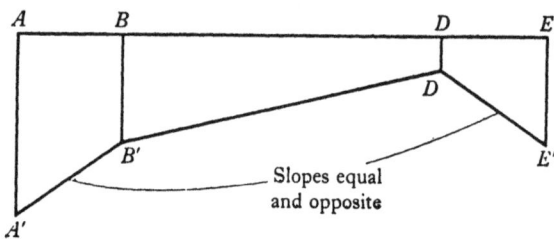

Fig. 2.54

consists of three straight lines, and the slopes of the lines for the two columns are equal and opposite. This last property follows immediately from the fact that the slope of a bending-moment diagram is equal to the shear force at that point, and the reactant shear forces in the two columns (T, see fig. 2.47(*b*)) are the same.

Reactant moments were expressed in fig. 2.48(*b*) in terms of certain actions introduced at a central cut, fig. 2.47(b). However, there is a wide choice of quantities that could be regarded as redundancies for this problem; whatever quantities are chosen, the two essential features of fig. 2.54 will be reproduced. The diagram will consist of three straight lines, and the column slopes will be equal and opposite. Figure 2.54 requires the knowledge of only three quantities in order that the diagram should be completely specified, that is, the diagram is determined once three values of the reactant moments are fixed. If the reactant moments at D and E are known, for example, then line $D'E'$ can be drawn. If, in addition, the reactant moment at A is specified, then the line $A'B'$ can be drawn of equal and opposite slope to the line $D'E'$, and the line $B'D'$ can then be completed.

Precisely this technique would give a quick graphical way of tackling the numerical problem just discussed. To draw by trial and error the reactant line in fig. 2.53(b), a value of M_p would first be assumed. This value could be set off from the free bending moments at D and E, so that $D'E'$ could be drawn, and at A, so that $A'B'$ could be drawn. A check would then be made that the largest bending moment elsewhere in the frame (at B_3 in this example) was equal to the assumed value of M_p; if not, a new value of M_p would be guessed and the process repeated.

Before leaving this particular numerical example, it may be noted that although the calculations (of free and reactant moments) have been made for a fixed-base frame, the corresponding pin-based frame can be investigated with almost no extra labour. This may be of practical importance if a designer has to balance the cost of the stronger foundations required for a fixed-base frame against the extra material required for a pin-based frame to carry the same loads. All that is necessary is that the free and reactant diagrams of figs. 2.51 and 2.54 should be superimposed in such a way that there is no net bending moment at the column feet.

Using trial and error in the way just described, the collapse bending-moment diagram of fig. 2.55 is quickly arrived at, with hinges formed at B_2 and D in addition to the pinned bases at A and E. This mode of collapse can be confirmed by writing the collapse equations:

$$\left.\begin{aligned}
A: &\quad 96-(M+12S+8T) = 0,\\
B_2: &\quad 12-(M+\ 6S) \qquad = -M_p,\\
D: &\quad 48-(M-12S) \qquad = M_p,\\
E: &\quad 0-(M-12S+8T) = 0.
\end{aligned}\right\} \qquad (2.57)$$

The zeros on the right-hand sides of the equations for the pinned feet at A and E should be noted. The equations solve to give

$$\left.\begin{array}{r} M_p = 54, \\ M = 42, \\ 12S = 48, \\ 8T = 6. \end{array}\right\} \tag{2.58}$$

and table 2.7 gives the bending moments corresponding to fig. 2.55.

Table 2.7

	A	A_1	B	B_1	B_2	B_3	C	D_3	D_2	D_1	D	E_1	E
Free	96	60	48	27	12	3	0	3	12	27	48	36	0
Reactant	−96	−93	−90	−78	−66	−54	−42	−30	−18	−6	6	3	0
Total	0	−33	−42	−51	−54	−51	−42	−27	−6	21	54	39	0

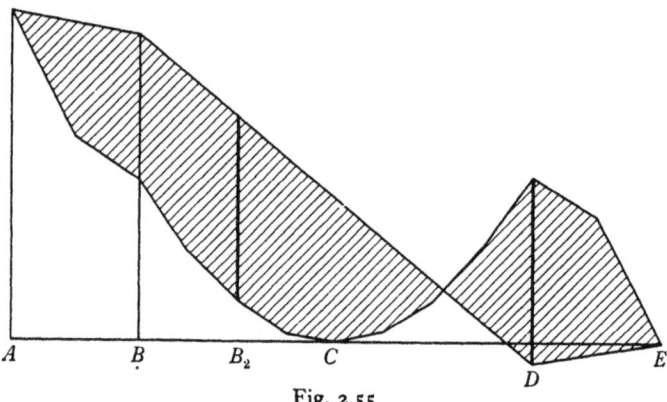

Fig. 2.55

2.8 Pitched-roof portal frames

Difficulties in the analysis of the pitched-roof portal frame lie almost entirely in the more complicated geometry compared with the rectangular frame; otherwise, the techniques given above can be used without change. The fixed-base frame of fig. 2.56 carries the idealized loads shown; the three basic modes of collapse, corresponding to those of fig. 2.43 for the rectangular frame, are sketched in fig. 2.57. Confining the discussion to a frame of uniform section, full plastic moment M_p, it will be seen that the pure sidesway mode of fig. 2.57(a) occurs for

$$Hh_1 = 4M_p. \tag{2.59}$$

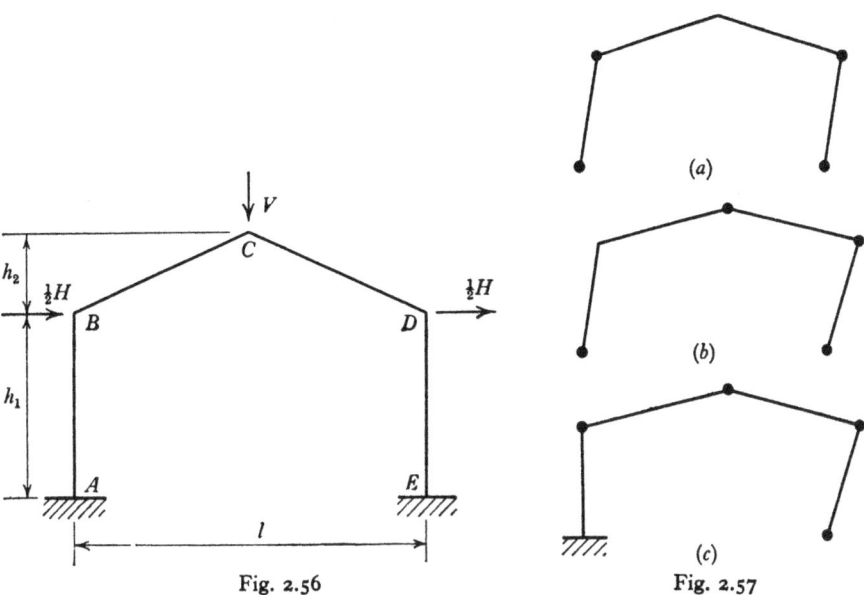

Fig. 2.56   Fig. 2.57

This expression may be compared with the first of (2.42) for the corresponding rectangular frame.

Mode (c) of fig. 2.57 is of the greatest practical interest; it corresponds to collapse of the beam only in the rectangular frame, and is the most usual mode for real loading (although the hinge at C may be displaced slightly from the apex). In fig. 2.58, the instantaneous centre I_{CD} of the

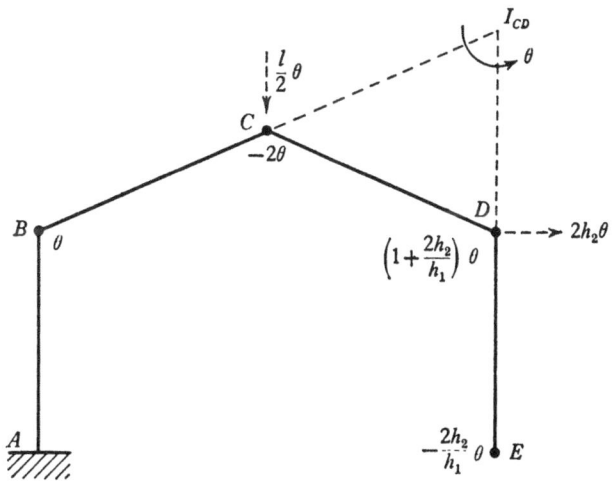

Fig. 2.58

99

rafter CD is located at the intersection of BC and ED produced; this follows from the fact that the motion of C must be perpendicular to BC, since B is a fixed point, and, similarly, D moves horizontally perpendicular to ED. It is convenient to specify the motion of the mechanism in terms of an infinitesimal rotation θ about the instantaneous centre I_{CD}; thus the rafter CD will rotate through an angle θ from its original position. Since the *horizontal* distance between I_{CD} and C is $\frac{1}{2}l$, the vertical deflexion of C is simply $\frac{1}{2}l\theta$. However, the horizontal distance between B and C is also $\frac{1}{2}l$, so that the hinge rotation at B must be θ, and rafter BC will make an angle θ with its original direction. Since both rafters rotate through θ, in opposite senses, the hinge rotation at C must be 2θ.

Similar considerations establish the values of the rotations at hinges D and E. The distance DI_{CD} is $2h_2$, so that D moves horizontally through a distance $2h_2\theta$. Thus the numerical value of the hinge rotation at the column foot E must be $2h_2\theta/h_1$, and finally the value of the rotation at D marked in fig. 2.58 may be deduced.

Having established these deflexions and rotations, the work equation gives immediately

$$\tfrac{1}{2}Vl + Hh_2 = \left(4 + 4\frac{h_2}{h_1}\right) M_p, \tag{2.60}$$

and this may be compared with the last of equations (2.42) for the case $h_2 = 0$. The pitch of the frame in fig. 2.58 ensures that at least one of the columns must participate in the collapse mechanism, and equation (2.60) indicates the strengthening effect, namely a factor

$$(1 + h_2/h_1).$$

For the collapse mode of fig. 2.58 to be correct, the bending moment at the column foot A must be less numerically than M_p. The shear balance of fig. 2.59 leads to the equation

$$Hh_1 = M_p + M_A, \tag{2.61}$$

so that
$$-M_p \leqslant Hh_1 - M_p \leqslant M_p,$$

or
$$0 \leqslant Hh_1 \leqslant 2M_p. \tag{2.62}$$

The value of M_p may be introduced from equation (2.60), and the condition then becomes

$$0 \leqslant Hh_1(1 + \tfrac{1}{2}h_2/h_1) \leqslant \tfrac{1}{4}Vl. \tag{2.63}$$

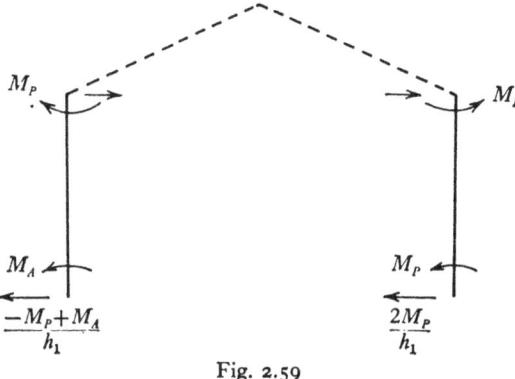

Fig. 2.59

Mode (b) of fig. 2.57 may be analysed in the same way by again considering a rotation θ about the instantaneous centre I_{CD}, which now lies on the intersection of AC and ED produced. The collapse equation is

$$\tfrac{1}{2}Vl + Hh_1\left(1 + 2\frac{h_2}{h_1}\right) = \left(6 + 4\frac{h_2}{h_1}\right)M_p. \qquad (2.64)$$

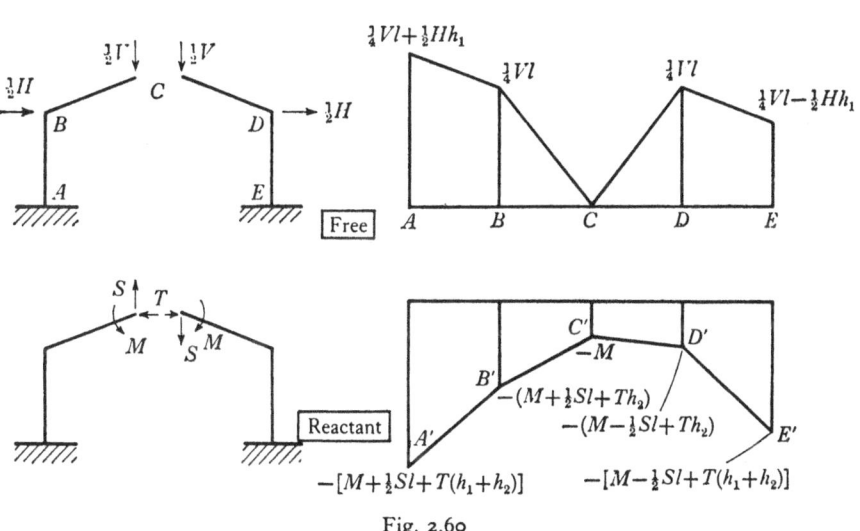

Fig. 2.60

Bending-moment diagrams may be constructed as usual from free and reactant systems, obtained as for the rectangular frame by cutting at the centre of span C. Figure 2.60 shows the free and reactant diagrams with the leading values marked. The reactant line is again specified in terms of three unknown quantities, M, S and T, but now consists of four straight lines, the pitch of the roof being reflected in the portion $B'C'D'$ of the diagram.

Mode (c) of fig. 2.57 may be analysed again by writing the collapse equations:

$$
\left.\begin{aligned}
B: \ (\tfrac{1}{4}Vl) & \quad -(M+\tfrac{1}{2}Sl+Th_2) & = M_p, \\
C: \ (\text{o}) & \quad -(M) & = -M_p, \\
D: \ (\tfrac{1}{4}Vl) & \quad -(M-\tfrac{1}{2}Sl+Th_2) & = M_p, \\
E: \ (\tfrac{1}{4}Vl-\tfrac{1}{2}Hh_1) & -(M-\tfrac{1}{2}Sl+Th_1+Th_2) & = -M_p,
\end{aligned}\right\} \quad (2.65)
$$

from which equation (2.60) may be derived by eliminating M, S and T. The values of these quantities are

$$
\left.\begin{aligned}
M &= M_p, \\
S &= \text{o}, \\
T(h_1+h_2) &= \tfrac{1}{4}Vl-\tfrac{1}{2}Hh_1.
\end{aligned}\right\} \quad (2.66)
$$

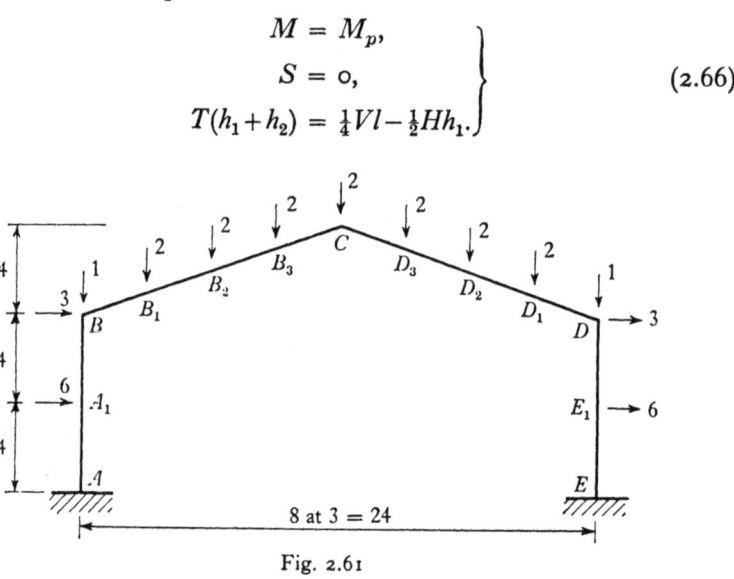

Fig. 2.61

The numerical example of fig. 2.50 will be repeated for a pitched-roof frame, fig. 2.61. With the loads and dimensions chosen, the free bending-moment diagram is identical with that of fig. 2.51 for the rectangular frame. The reactant line will be as shown in fig. 2.60, with $\tfrac{1}{2}l = 12$, $h_1 = 8$ and $h_2 = 4$. Thus, assuming collapse to occur by mode (c) of fig. 2.57,

$$
\left.\begin{aligned}
B: \ & 48-(M+12S+\ 4T) = M_p. \\
C: \ & \text{o}-(M) = -M_p, \\
D: \ & 48-(M-12S+\ 4T) = M_p, \\
E: \ & \text{o}-(M-12S+12T) = -M_p.
\end{aligned}\right\} \quad (2.67)
$$

These equations solve to give

$$M_p = M = 24, \\ S = T = 0, \Big\} \qquad (2.68)$$

and table 2.8 displays the corresponding bending moments. The bending moment at column foot A seriously violates the yield condition, and mode (b) of fig. 2.57 will next be tried, involving hinges at A, C, D and E:

$$
\begin{aligned}
A: & \quad 96 - (M + 12S + 12T) = M_p, \\
C: & \quad 0 - (M) = -M_p, \\
D: & \quad 48 - (M - 12S + 4T) = M_p, \\
E: & \quad 0 - (M - 12S + 12T) = -M_p.
\end{aligned}
\Bigg\} \qquad (2.69)
$$

Table 2.8

	A	A_1	B	B_1	B_2	B_3	C	D_3	D_2	D_1	D	E_1	E
Free	96	60	48	27	12	3	0	3	12	27	48	36	0
Reactant	−24	−24	−24	−24	−24	−24	−24	−24	−24	−24	−24	−24	−24
Total	72	36	24	3	−12	−21	−24	−21	−12	3	24	12	−24

Table 2.9

	A	A_1	B	B_1	B_2	B_3	C	D_3	D_2	D_1	D	E_1	E
Free	96	60	48	27	12	3	0	3	12	27	48	36	0
Reactant	−66	−60	−54	−48	−42	−36	−30	−27	−24	−21	−18	−24	−30
Total	30	0	−6	−21	−30	−33	−30	−24	−12	6	30	12	−30

These equations solve to give

$$M_p = M = 30, \\ 12S = 12T = 18, \Big\} \qquad (2.70)$$

and table 2.9 displays the new set of bending moments.

The solution is now almost complete; the correct mode of collapse is that sketched in fig. 2.62, for which

$$
\begin{aligned}
A: & \quad 96 - (M + 12S + 12T) = M_p, \\
B_3: & \quad 3 - (M + 3S + T) = -M_p, \\
D: & \quad 48 - (M - 12S + 4T) = M_p, \\
E: & \quad 0 - (M - 12S + 12T) = -M_p.
\end{aligned}
\Bigg\} \qquad (2.71)
$$

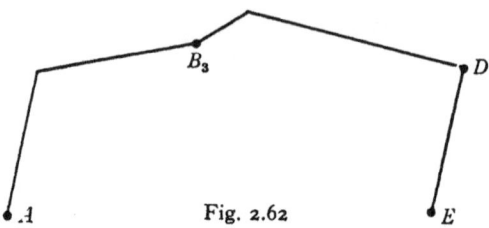

Fig. 2.62

Table 2.10

	A	A_1	B	B_1	B_2	B_3	C
Free	96	60	48	27	12	3	o
Reactant	−65·25	−58·5	−51·75	−45·75	−39·75	−33·75	−27·75
Total	**30·75**	1·5	−3·75	−18·75	−27·75	**−30·75**	−27·75

	D_3	D_2	D_1	D	E_1	E
Free	3	12	27	48	36	o
Reactant	−25·1	−22·5	−19·9	−17·25	−24	−30·75
Total	−22·1	−10·5	7·1	**30·75**	12	**−30·75**

From these equations,

$$M_p = 30\cdot75,$$
$$M = 27\cdot75,$$
$$12S = 17\cdot25,$$
$$12T = 20\cdot25,$$

(2.72)

and table 2.10 confirms that the collapse mode is correct. The final bending-moment diagram is sketched in fig. 2.63.

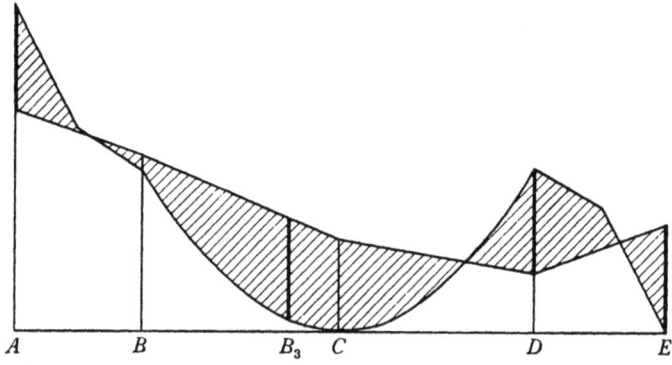

Fig. 2.63

104

The final value of M_p, 30·75, is only marginally less than the corresponding value (32·3) for the rectangular frame, table 2.6. The strengthening effect of the pitched roof is small for this case of relatively high wind load.

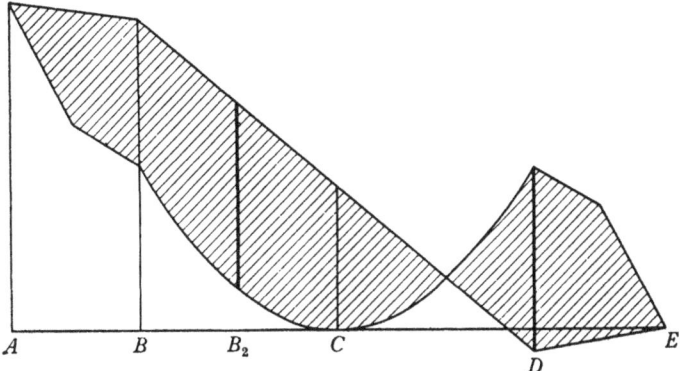

Fig. 2.64

Table 2.11

	A	A_1	B	B_1	B_2	B_3	C
Free	96	60	48	27	12	3	0
Reactant	−96	−93·3	−90·7	−78	−65·3	−52·7	−40
Total	0	−33·3	−42·7	−51	−53·3	−49·7	−40

	D_3	D_2	D_1	D	E_1	E
Free	3	12	27	48	36	0
Reactant	−28·7	−17·3	−6	5·3	2·7	0
Total	−25·7	−5·3	21	53·3	38·7	0

As before, the pin-based frame may be analysed immediately, by writing the condition that the total bending moment at both column feet is zero. Figure 2.64 shows the collapse bending-moment diagram, derived from the solution to the following equations:

$$\left. \begin{aligned} A: \quad & 96 - (M + 12S + 12T) = 0, \\ B_2: \quad & 12 - (M + 6S + 2T) = -M_p, \\ D: \quad & 48 - (M - 12S + 4T) = M_p, \\ E: \quad & 0 - (M - 12S + 12T) = 0. \end{aligned} \right\} \quad (2.73)$$

The resulting values ($M_p = 53 \cdot 3$, $M = 40$, $12S = 48$, $12T = 8$) have been used to calculate the bending moments of table 2.11.

EXAMPLES

2.1 A beam of uniform cross-section, full plastic moment M_p, of length $2l$ rests on simple supports at its ends and on a central prop. Equal concentrated loads are applied to the centre of each span. Show that collapse occurs when the loads have the value $6M_p/l$.

2.2 If the loads in example 2.1 were not applied at the centres of the spans but at a distance al from each end of the beam what would be their value at collapse?

$$\left(Ans.\ \frac{(1+a)}{a(1-a)}\,M_p/l. \right)$$

2.3 If the beam of example 2.1 carried four equal loads W, symmetrically arranged about the centre of length, one at the centre of each span and one at a distance $\frac{1}{8}l$ from each end of the beam, what would be the value of W at collapse? $\left(Ans.\ \frac{24}{5}M_p/l. \right)$

2.4 If the beam of example 2.1 carried four equal loads, W, symmetrically arranged about the centre of length, one at the centre of each span and one at a distance $3l/8$ from each end of the beam, what would be the value of W at collapse? $\left(Ans.\ \frac{88}{27}M_p/l. \right)$

2.5 If the beam of example 2.1 carried four equal loads W, symmetrically arranged about the centre of length, one at the centre of each span and one at a distance al from each end of the beam, what would be the value of a if at collapse the whole length of beam between the loads on each span became plastic? $\left(Ans.\ \frac{1}{4}. \right)$

2.6 If the beam of example 2.1 is subjected to a concentrated load at the centre of the left-hand span only, what wiil be the magnitude of that load at collapse? $\left(Ans.\ 4M_p/l. \right)$

2.7 A beam of uniform cross-section, full plastic moment M_p, of length $2l$ is encastré at its ends and also rests on a central prop. Equal concentrated loads are applied to the centre of each span. Show that collapse occurs when the loads have the value $8M_p/l$.

2.8 If the loads in example 2.7 were not applied at the centres of the spans but at a distance al from each of the encastré ends, what would be their value at collapse?

$$\left(Ans.\ \frac{2}{a(1-a)}\,M_p/l. \right)$$

2.9 If the beam of example 2.7 carried four equal loads W, symmetrically arranged about the central prop, one at the centre of each span and one at a distance $\frac{1}{3}l$ from each encastré end, what would be the value of W at collapse?
(*Ans* $\frac{32}{5}M_p/l$.)

2.10 If the beam of example 2.7 is subjected to a concentrated load at the centre of the left-hand span only, what will be the magnitude of that load at collapse? (*Ans.* $8M_p/l$.)

2.11 An encastré beam length l, of uniform section, full plastic moment M_p, is subjected to a uniformly distributed load W and to a concentrated load $0\cdot5W$ at a distance of $\frac{1}{3}l$ from one end. Find the value of W to cause collapse.
(*Ans.* $9M_p/l$.)

2.12 A cantilever of uniform section and length 3 m is propped at its end. It carries a uniformly distributed load of 30 kN/m. When, in addition, a concentrated load of 25 kN is applied, 1 m from the prop, the beam is on the point of collapse. Show that the full plastic moment of the beam is 35 kNm.

2.13 If the concentrated load of 25 kN had been applied 1 m from the encastré end of the propped cantilever of example 2.12, already carrying the uniformly distributed load 30 kN/m, what would have been the load factor? (*Ans.* 1·13.)

2.14 A beam of uniform section, full plastic moment M_p, length l, is built-in at one end and simply supported at the other. It carries a concentrated load W at a distance a from the built-in end. Show that at collapse W has the value

$$\frac{2l-a}{a(l-a)}M_p.$$

Show that if both ends had been built-in the load at collapse would have increased in the ratio
$$\frac{2l}{2l-a}.$$

2.15 A beam of uniform section, full plastic moment M_p, length l, was built in at the left-hand end A, and supported on a prop at the right-hand end B, so that before loading the beam was horizontal. When a uniformly distributed load was applied some settlement of the end A occurred until, when the total load was $8M_p/l$, the restraining moment at A was zero. Find the position of the first plastic hinge and the total load which ultimately produced collapse, assuming that no further settlement occurred.
(*Ans.* Centre of length: $2(3+2\sqrt{2})M_p/l$.)

2.16 A beam $ABCD$ of uniform section throughout, full plastic moment M_p, is pinned to four supports so forming a continuous beam of three equal spans, length l. A load W is applied at the centre of each span. Find the value of W which causes collapse. (*Ans.* $6M_p/l$.)

2. SIMPLE BEAMS AND FRAMES

2.17 If only one load W had been applied to the centre of the continuous beam of example 2.16, what would have been its value at collapse? If the outer spans had been twice as long as the centre, what would have been the effect on the value of the collapse load? *(Ans. $8M_p/l$.)*

2.18 The continuous beam shown has three equal spans carrying central point loads. There is no change of beam section between supports but the plastic moment of resistance of the outer spans is only two-thirds that of the central span which is M_p. At what value of W does collapse occur? *(Ans. $\frac{20}{3}M_p/l$.)*

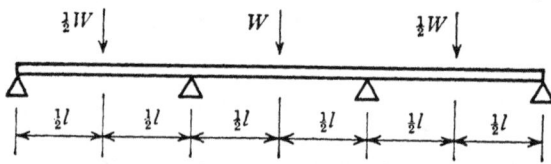

2.19 A uniform continuous beam, of full plastic moment M_p, rests on five simple supports A, B, C, D and E. $AB = 6l$, $BC = CD = 8l$, $DE = 10l$. Each span carries a concentrated load at its midpoint, these loads being W on AB, W on BC, $1\cdot4W$ on CD and $0\cdot5W$ on DE. Find the value of W which will just cause collapse. *(Ans. $5M_p/7l$.)*

2.20 A continuous beam rests on four supports A, B, C and D; $AB = BC = CD = 4$ m. Each span carries a uniformly distributed load: 25 kN on AB, 50 kN on BC and 30 kN on CD. The beam is to be of uniform section between supports but the section of the centre span is to be heavier than that of either outer span. Find the required value of the full plastic moment for each span if a load factor of 2 at collapse is to be provided. *(Ans. 17·2, 31·1, 20·6 kNm.)*

2.21 The three-span continuous beam shown is to be made of three steel members each of uniform section throughout. If, under the given loads, collapse is to occur in all spans simultaneously, show that possible values for the full plastic moments of the members are 58·3, 25 and 38·3 kNm.

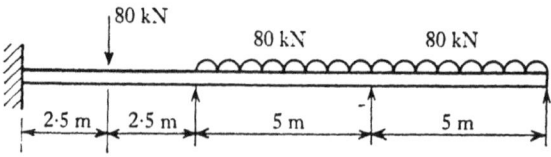

2.22 The continuous beam shown was of uniform section throughout, and collapsed when subjected to the given loads. What was the full plastic moment of the beam? *(Ans. 255 kNm.)*

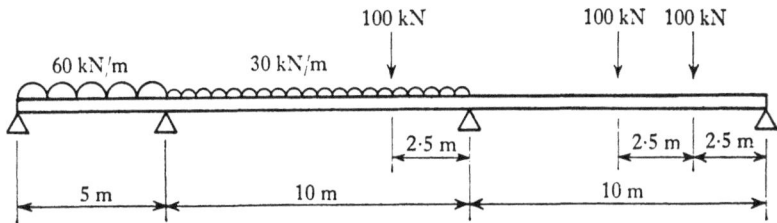

2.23 If, in example 2.16, the loads W are each 300 kN and the span l is 6 m, show that a possible design for the beam, in a mild steel with a yield stress $\sigma_0 = 250$ N/mm², is a 16 × 6 UB 67 kg.

2.24 A beam of uniform section, 10 m long, is to be part of the floor system in a warehouse. It is simply supported at its ends and rests on a central support. The dead load and superimposed working live load together amount to a uniformly distributed load of 65 kN/m. A load factor of 1·75 is to be provided. Find a suitable section for the beam in a steel having $\sigma_0 = 250$ N/mm². *(Ans.* 16 × 7 UB 54 kg.)

2.25 The beam ABC shown is to be designed to carry the given loads with a load factor of 1·75. Using a steel with a minimum yield stress of 250 N/mm² show that a possible design for the beam is a 21 × 8¼ UB 92 kg, continuous from A to C.

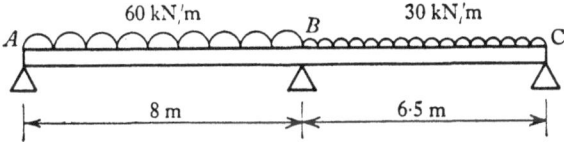

This design is extravagant in weight. It is decided therefore to use a smaller I-section continuous from A to C and to add symmetrical flange reinforcement where necessary in span AB. Show that a basic section 14 × 6¾ UB 45 kg will be satisfactory, and find the size of the flange plates that must be added and their net length. *(Ans.* 300 × 20 mm; 6·5 m.)

2.26 The three-span continuous beam shown carries a working load of 48 kN/m. If a load factor of 1·75 is to be provided by a beam of uniform section in a steel with yield stress 250 N/mm², show that an 18 × 6 UB 74 kg will be needed.

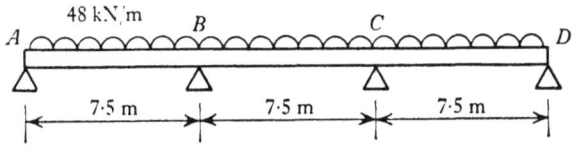

A plated beam would be used more efficiently over span *BC*. Show that a basic section 16×7 UB 60 kg. would be satisfactory, and find the sizes of the flange plates required.   (*Ans.* 190×8 mm.)

2.27   Loads of 420 kN each are to be supported as shown at the centres of the spans of a continuous beam, with a load factor of 1·75. For architectural reasons the overall depth of the beam is limited to 0·5 m. Show that, using steel with $\sigma_0 = 250$ N/mm², an $18 \times 7\frac{1}{2}$ UB 98 kg is suitable, plated in the regions under the loads. Find the sizes and lengths of the flange plates.

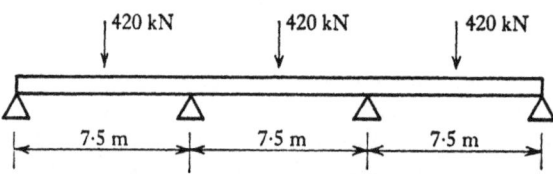

(*Ans.* 220 mm × 10 mm plates, 1·44 m long in centre span,
280 mm × 16 mm plates, 3·08 m long in end spans.)

2.28   A load of 100 kN/m is to be carried over the three spans shown with a load factor of 1·75. It is decided to use an I-section, in steel with $\sigma_0 = 250$ N/mm², running continuously over all spans with added flange plates

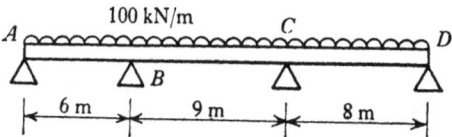

running continuously over support *C* into the spans *BC* and *CD*. Find the size of basic section and length and area of flange plates required.

(*Ans.* $21 \times 8\frac{1}{4}$ UB 92 kg; 200 mm × 15 mm, 13·8 m.)

2.29   A continuous beam *ABC* rests on simple supports at *A*, *B* and *C*, the span *AB* being *l* and the span *BC* being $2l$. The beam is composed of two uniform members *AB* and *BC* rigidly jointed at *B*, the member *AB* having a fully plastic moment M_p and the member *BC* having a fully plastic moment KM_p. The span *AB* carries a central concentrated load *W*, and the span *BC* carries a uniformly distributed load $\frac{8}{8}W$.

The beam is to be designed according to the plastic theory, so that for any given value of *K* the value of M_p is to be such that one or other of the spans fails by plastic collapse. Show that collapse will occur in the span *AB* rather than in the span *BC* if *K* exceeds a value of about 0·8.

It is intended to design the beam ABC so that the total weight of material used is as small as possible. It may be assumed that there is a continuous range of sections available, and that the weight of each member per unit length is proportional to its fully plastic moment. Sketch a curve showing how the total weight of material in the beam varies with K for values of K between 0·5 and 2, and show that the minimum weight of material which can be used is about $0·46\lambda Wl^2$, where λ is the weight per unit length of a member with a fully plastic moment of unity. (*M.S.T.* II, 1952.)

2.30 The propped cantilever shown is fabricated from two 300×25 mm flange plates, and a thin web plate which may be neglected in determining the plastic moment of resistance. The web depth tapers uniformly from 750 to

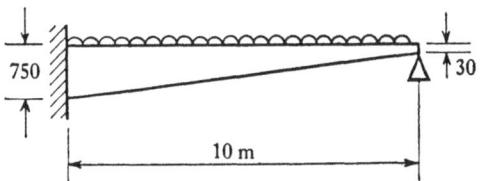

750

30

10 m

30 mm. The yield stress of the flanges is 250 N/mm². Determine the uniformly distributed collapse load. (Imperial College 1964, adapted.)

(*Ans.* 820 kN.)

2.31 A propped cantilever of uniform section, length l, full plastic moment M_p, flexural rigidity EI, is initially stress free. It is subjected to a uniformly distributed load which is gradually increased until collapse occurs. Assuming that the moment/curvature relationship is that of fig. 2.1, determine the deflexion at incipient collapse at the cross-section where the final hinge forms.

(*Ans.* $0·089 M_p l^2/EI$.)

2.32 An I-beam, having thin flanges and web, is simply supported and subjected to a uniformly distributed load of intensity two-thirds of that which would cause collapse. A rigid prop is inserted at the centre and the load is increased until collapse occurs. Trace the development of the plastic hinges which will be formed. The effects of shear and instability are to be neglected. (*M.S.T.* II, 1950.)

2.33 A steel I-joist is designed to carry, as a simply-supported beam, a uniformly distributed load of intensity w over a span $2l$ with a load factor of 2.

When a load of intensity $2w$ has been applied, so that the beam is about to collapse, a rigid prop is inserted at the centre. Show that more load can be added until yield occurs at sections $3l/5$ from each end. Show further that complete collapse of the propped beam does not take place until the total intensity of load carried is $2(3 + 2\sqrt{2})w$.

It is to be assumed that the thickness of the flanges and web are negligible in comparison with the depth of the member and that no buckling of the web or other condition of instability occurs. (*M.S.T.* II, 1947.)

2.34 An encastré beam of span L is to be designed for collapse under a single central point load λW. For a distance $\frac{1}{2}kL$ at each end of the span the fully plastic moment is to have r times the value obtaining for the remainder of the span. The weight per unit length of beam is

$$\beta \left[1 + \frac{32M}{\lambda WL} \right],$$

where M is the fully plastic moment at the section in question, and β is a constant.

Determine the ratios r and k for minimum weight of the beam, and find the saving in material as compared with a design using a uniform beam. (University of London B.Sc. (Engineering) Part III: Civil, King's College, 1965.)

(*Ans. $r = 5/3$, $k = 1/4$, 10%.*)

2.35 In a fixed-base rectangular portal frame $ABCD$ the column AB is of height 16 and the column DC is of height 24; the beam BC of length 16 is horizontal. All the members of the frame have the same full plastic moment M_p. The beam BC carries a central concentrated vertical load of 70, and a concentrated horizontal load of 28 is applied at C in the direction BC. Find the value of M_p so that collapse just occurs. (*Ans.* 189.)

2.36 In example 2.35, the horizontal load of 28 is reversed in direction. Find M_p. (*Ans.* 176.)

2.37 In example 2.35, the frame is not uniform, but the columns have the same full plastic moment M_1 and the beam has full plastic moment M_2. Find M_1 and M_2 so that collapse just occurs whichever way the horizontal load acts. (*Ans.* 134·4, 280.)

2.38 A fixed-base rectangular portal frame is of height and span l. The columns each have full plastic moment $2M_p$, and the beam has full plastic moment M_p. One of the columns is subjected to a uniformly distributed horizontal load W. Find the value of W which would cause collapse.

(Answer is too strong a hint.)

2.39 Repeat example 2.38 for a uniform frame of full plastic moment M_p.

(*Ans.* $2(2+\sqrt{3})M_p/l$.)

2.40 The rigid frame shown has 6×6 UC 37 kg columns and a $10 \times 5\frac{3}{4}$ UB 43 kg beam; the yield stress of the steel is 250 N/mm². Neglecting the effects of axial load and instability, determine the ratio of W to H for all possible modes of plastic collapse of the frame. Determine the value of W when $W = 1·8H$. (Institution of Structural Engineers, 1962; adapted.)

(*Ans.* 63·6 kN.)

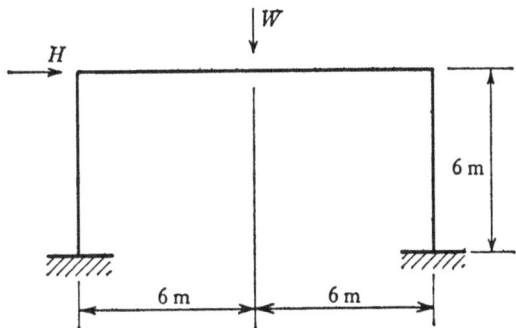

2.41 In the figure the load factor, λ, and the plastic moment, M, are to be regarded as variables, as well as W, P, L and H. Plot, in the plane of the dimensionless variables $(\lambda WL/M)$ and $(\lambda PH/M)$, the collapse locus for the family of loaded frames represented.

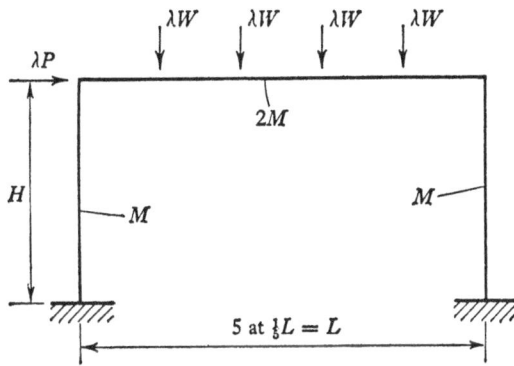

A frame conforming to the figure has $H:L = 4:3$. It is designed to a load factor of 2 under vertical loading only. What is the limiting value of $P:W$ for this frame if under combined loading the load factor may be reduced to 3/2? (University of London B.Sc. (Engineering) Part III: Civil, King's College, 1965.) (*Ans.* 13:20.)

2.42 A single storey building consists of a number of frames, each frame being comprised of several equal bays of pitched-roof portals rigidly connected at all joints. Considering one frame subjected only to a uniformly distributed vertical dead load on the roof, show that collapse of the frame will occur only in the bays at either end, providing that the rafter section is continuous and uniform throughout, and that the two end stanchions have equal sections (different from that of the rafters).

2.43 The frame in example 2.42 is subjected to a wind load, in addition to the dead load. The effect of the wind is to produce a small uniform pressure on the windward stanchion, and an equal suction on the leeward stanchion; all other wind effects are to be ignored. Show that collapse of the frame occurs only in the leeward bay.

2.44 The pitched roof portal frame shown is of the same cross-section and ductile material throughout. All the joints are rigid and the feet are fixed in foundations which can develop the full plastic moment of the members. Loads of the magnitudes shown are uniformly distributed over the members.

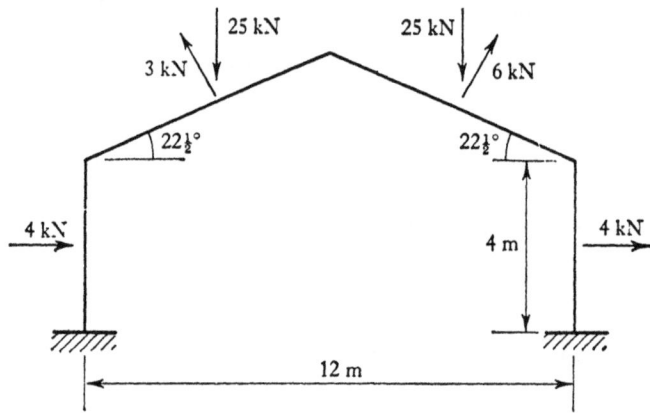

Estimate the required value of the full plastic moment so that the load factor against plastic collapse shall be 1·4. Instability effects and the reduction of plastic moments due to axial forces are to be neglected. (*M.S.T.* II, 1953; adapted.) (*Ans.* 31·5 kNm.)

2.45 The portal frame shown is of uniform section throughout. It is to be designed to resist
(i) dead plus superimposed loading, load factor 1·75;
(ii) dead plus superimposed plus wind loading, load factor 1·4. The loading is as shown, the dead plus superimposed loading being indicated

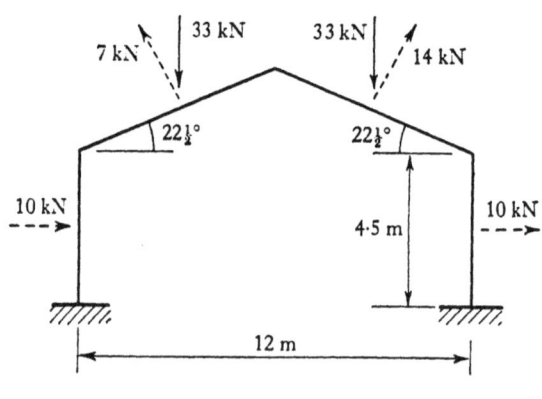

by the full arrows and the wind loading by the dotted arrows. All the loads may be regarded as uniformly distributed over the members. Find the required value of M_p. (*Ans.* 57·7 kNm.)

2.46 The continuous symmetrical pitched roof portal frame shown has rigid joints and is rigidly fixed to foundations at A and E. The frame has a uniform full plastic moment of M_p, on which the effect of axial loads may be neglected.

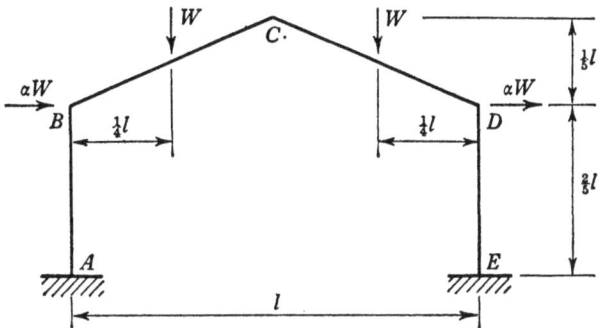

Investigate the value of W at which collapse will just occur for the load system shown. Present the solution in such a way that the mode of collapse and the value of W at collapse are indicated for any value of α. Neglect the effect of the deformation of the structure on the equations of equilibrium. (*M.S.T.* II, 1963.)

2.47 A symmetrical pitched-roof portal has a span of 7·2 m. The height at the ridge is 5·4 m and at the eaves 3·6 m. All joints are rigid and the feet of the stanchions are encastré. The stanchions have a fully plastic moment M_p and the roof members αM_p. The portal is subjected to a vertical load of 20 kN at the ridge and an outwardly directed horizontal load in the plane of the portal of 10 kN at the top of one stanchion.

For a given value of α find the least value of M_p which will prevent collapse and plot this value of M_p against α. (*M.S.T.* II, 1954; adapted.)

2.48 The plane portal frame with rigid joints shown is made of steel joist with a full plastic moment of resistance about the axis of bending of 12 kNm.

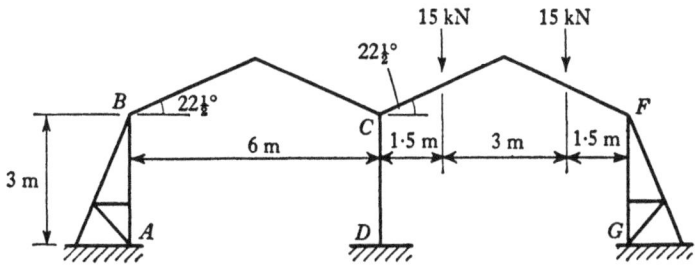

2. SIMPLE BEAMS AND FRAMES

The outer stanchions *AB* and *FG* are rigidly supported and the foot of the centre stanchion *CD* is encastré. By means of the plastic theory, making the usual simplifying assumptions, estimate the factor by which the loading shown would have to be multiplied for collapse just to occur. (*M.S.T.* II, 1958; adapted.) (*Ans.* 1·69.)

2.49 The north-light portal frame shown is composed of uniform members having full plastic moment 90 kNm, and has fixed feet and full strength joints; loads *V* and *H* are applied as shown. Plot a graph (interaction diagram) from which may be read the positive values of *V* and *H* which will just cause collapse of the frame. The usual assumptions of simple plastic theory may be made. (*M.S.T.* II, 1966; adapted.)

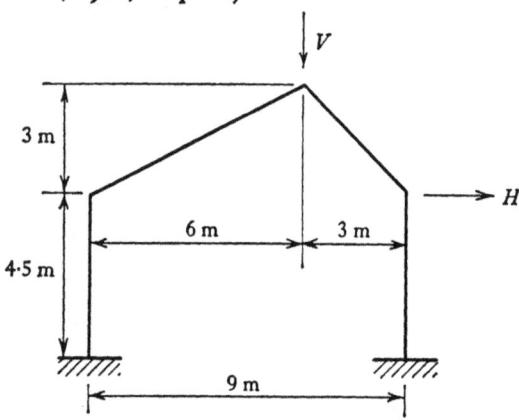

2.50 A pitched portal with hinged feet is subjected to vertical loading of constant intensity ω per unit horizontal distance, as indicated. The frame may be taken to have a plastic moment of resistance *M* throughout. Instability, and the effects on the plastic moment of axial and shear forces may be ignored.

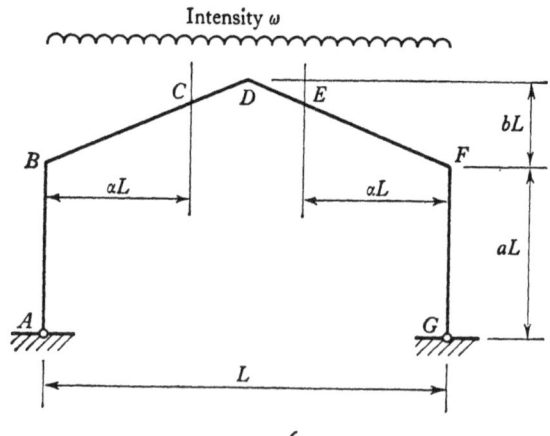

Show that when

$$\omega = \frac{4M(1+Q\alpha)}{L^2\alpha(1-\alpha)},$$

where $Q = b/a$, plastic collapse occurs with plastic hinges formed at B, F, C and E, the last two points being defined by

$$\alpha = \frac{1}{Q}\{\sqrt{(1+Q)}-1\}.$$

(University of London B.Sc. (Engineering) Part III: Civil, King's College, 1964.)

2.51 The frame of uniform section shown is to be designed by simple plastic theory to carry the uniformly distributed load W. Two designs are made, one for a frame with pinned feet, and one for fixed feet. Show that the ratio of full plastic moments for the two designs is

$$\left[\frac{1+\sqrt{(1+2k)}}{1+\sqrt{(1+k)}}\right]^2 \quad (M.S.T.\ \text{II}, 1965.)$$

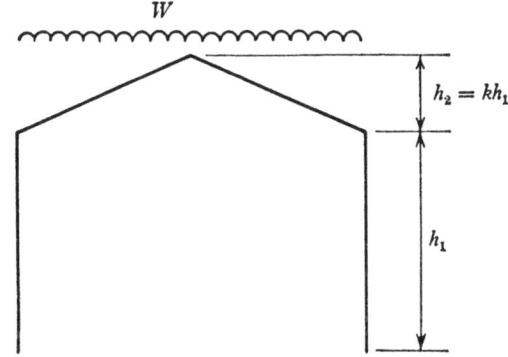

2.52 The two-bay pitched-roof portal frame shown has members of uniform section of full plastic moment M_p. The rafters are connected to each other and

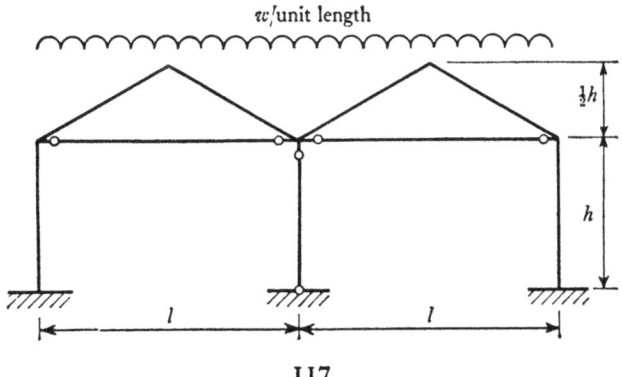

to the two external columns by full strength connexions. The external columns can sustain the full plastic moment at their feet, but the central column is a simple prop, pinned to the ground at one end and to the frame at the other. Light ties are connected across the frames at eaves level, and these are of sufficient section to prevent significant spread of the eaves when the frame is subjected to the factored collapse design load λw per unit length, uniformly distributed.

One of the ties is cut. Show that the load factor against collapse drops to 0.364λ.

The usual assumptions of simple plastic theory may be made. (*M.S.T.* II, 1961.)

2.53 The figure shows part of a continuous multi-bay frame of pitched-roof construction in which all the columns are rigidly fixed to the foundations. The dimensions of all bays are the same, the span of each being $5h$, the height to eaves h and the rise of the rafters h. The rafters BC and CD of the outer bay

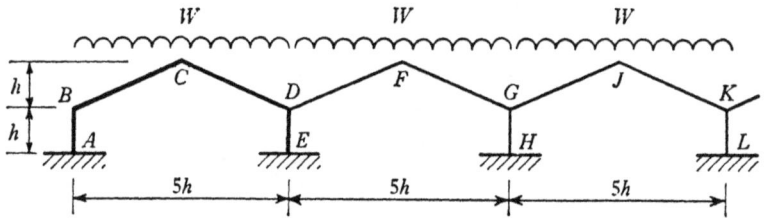

and the columns AB and DE have the same uniform full plastic moment M_1, while the inner rafters DF, FG, GJ, JK, etc., and the columns GH, KL, etc., have a smaller full plastic moment M_2. Each bay sustains a total vertical load W uniformly distributed. In designing the frame to a load factor at plastic collapse of 1.75, the value of M_2 is first chosen, and is made as small as possible. The necessary value of M_1 is then calculated. Obtain expressions for M_1 and M_2, making the approximation that, in the outer bay $ABCDE$, plastic hinges are confined to the joints and foundations.

The effect of elastic deformations on the collapse load and the effect of axial loads on the values of the full plastic moments are to be ignored. (*M.S.T.* II, 1960.) (*Ans.* $M_1 = 1.75(\frac{23}{128}Wh)$, $M_2 = 1.75(\frac{5}{64}Wh)$.)

2.54 The portal frame shown is composed of three straight uniform members, flexural rigidity 4.7 MNm², full plastic moment 66.0 kNm. The base connexions at A and D are rigid; the connexions between the members at B and C are flexible, the moment–rotation characteristic of each connexion being given by the following table:

Moment (kNm)	12·5	25·0	37·5	50·0	62·5	66·0
Rotation (radians)	0·0023	0·0053	0·0091	0·0136	0·0189	0·0205

The portal is subjected to a uniformly distributed vertical load W acting over the full length of the beam. Calculate the central deflexion of the beam as W is increased slowly from zero up to the collapse value, and compare the resulting load-deflexion curve with that for a similar frame with rigid connexions at B and C.

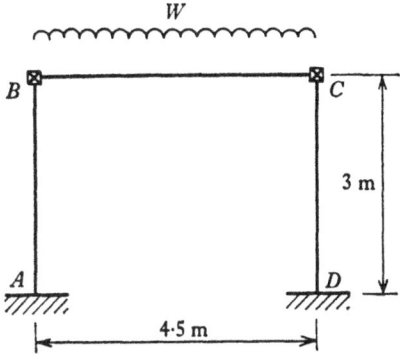

It may be assumed that a cross-section where the bending moment is less than 66·0 kNm remains completely elastic, and that indefinitely large rotations can occur at constant moment at a plastic hinge.

Deflexions need be calculated only at the six loading conditions corresponding to the six sets of values given in the table. (*M.S.T.* II, 1961; adapted.)

3

THE TOOLS OF PLASTIC DESIGN

The beams and portal frames discussed so far have been simple enough for them to be analysed by a 'direct' approach. Bending-moment diagrams were easy to draw, and the number of possible mechanisms of collapse was small, so that the complete behaviour of the frames in the various examples could be visualized readily. As the structure under examination becomes more complex, however, the difficulties of the direct approach increase very rapidly. The choice of collapse mechanisms is wide, and it becomes almost impossible to derive the correct solution immediately. In this situation it is natural to seek on the one hand simple approximate methods of analysis, and on the other some general principles against which the accuracy of the approximate methods can be measured.

The basic theorems of plastic theory will be established in this chapter, and, in setting up these basic theorems, an indication is given of ways in which approximate solutions may be derived. These, in turn, lead on to powerful techniques of practical computation, one of which, that of the combination of mechanisms, is discussed in detail in the next chapter. In establishing the principles of plastic theory the equation of virtual work is used extensively.

3.1 The equation of virtual work

The principle of virtual work is so simple as to be almost 'self-evident' but some surprising techniques of calculation can result from its use. Stated simply, if a body in equilibrium is given a set of small displacements, then the work done by the external loads on the external displacements is equal to the work done by the internal forces on the internal displacements. The following points may be noted:

(i) The system of displacements must be compatible, in the sense that the internal deformations must correspond to the external displacements. However, the deformations need not be 'real', that is, they need not correspond to any actual or possible state of equilibrium; the body can be arbitrarily distorted without reference to any loading system. This explains the use of the word 'virtual'.

(ii) The internal forces must be in equilibrium with the external loads. However, the internal forces need not be the actual forces due to those loads; any equilibrium set of forces may be used in the equation of virtual work.

For the present purposes, only the mechanism type of deformation will be considered. Hence all the internal deformations of the frame will be concentrated at hinges, which are connected together by members which remain straight; the hinge rotations will lead to certain displacements of the joints and loading points.

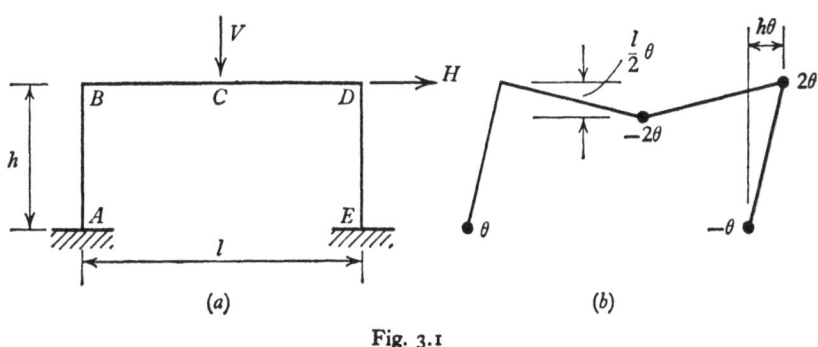

Fig. 3.1

Suppose, then, that certain bending moments M_i (e.g. M_A, M_B, M_C, M_D and M_E for the fixed-base portal frame, fig. 3.1(a)), are in equilibrium with certain external loads W_j on the frame (e.g. V and H in fig. 3.1(a)). The equilibrium statement is then

Loads W_j are in equilibrium with bending moments M_i,   (3.1)

or, more shortly, (W_j, M_i) is an equilibrium set. That is, for the portal frame of fig. 3.1(a), the five bending moments M_A to M_E satisfy every test of statical equilibrium, but are not necessarily the actual bending moments in the frame. Since the frame has three redundancies, the five bending moments can be written in general terms of three unknown quantities.

Quite independently of this equilibrium statement, suppose that certain hinge rotations ϕ_i in the frame lead to corresponding joint displacements δ_j. (For example, the mechanism of fig. 3.1(b) has hinge rotations $(\theta, 0, -2\theta, 2\theta, -\theta)$ at the sections A, B, C, D and E, and the displacements of the two loaded joints are $(\frac{1}{2}l\theta, h\theta)$. The signs of the hinge rotations correspond to the convention previously used; compression on the inner face of the frame is denoted positive.) The deformation statement is then

Joint displacements δ_j are compatible with hinge rotations ϕ_i, (3.2)

or, more shortly, the set (δ_j, ϕ_i) is compatible.

The equation of virtual work combines statements (3.1) and (3.2):

$$\Sigma \overset{\frown}{W_j \delta_j} = \Sigma \overset{\frown}{M_i \phi_i}.$$

(3.3)

As an example, equation (3.3) will be used to derive an equilibrium equation for the fixed-base rectangular frame, fig. 3.1(a). The bending moments round the frame are in equilibrium with the two loads V and H, so that statement (3.1) may be written in full:

(V, H) are in equilibrium with $(M_A, M_B, M_C, M_D, M_E)$. (3.4)

Taking (arbitrarily) the *virtual* mechanism of fig. 3.1(b), the following compatibility statement may be written:

$(\tfrac{1}{2}l\theta, h\theta)$ are compatible with $(\theta, 0, -2\theta, 2\theta, -\theta)$. (3.5)

Thus, combining (3.4) and (3.5) by means of the virtual work equation (3.3), and cancelling θ throughout,

$$\tfrac{1}{2}Vl + Hh = M_A - 2M_C + 2M_D - M_E.$$

(3.6)

This equation holds irrespective of whether the frame is elastic or plastic; no mention has been made of the behaviour of the material. Whatever the actual values of the bending moments at the sections A, C, D and E, they must always be related by equation (3.6).

If, in fact, the virtual mechanism of fig. 3.1(b), which furnished equation (3.6), is also a representation of the collapse mechanism, then the values of the bending moments at sections A, C, D and E are known to have the numerical value M_p. Since the signs of the bending moments must accord with the signs of the hinge rotations for a *real* mechanism,

$$M_A = M_D = M_p, \quad M_C = M_E = -M_p.$$

(3.7)

It will be seen that these particular values of the bending moments lead to the maximum possible value of the right-hand side of equation (3.6), if, numerically, each bending moment is to be at most equal to M_p. The resulting collapse equation,

$$\tfrac{1}{2}Vl + Hh = 6M_p$$

(3.8)

is identical with equation (2.42(b)), which was obtained directly by using the work equation.

This idea of a collapse equation being derivable from a more general equilibrium equation, indeed representing in some sense the breakdown of an equilibrium equation, will be developed further in the next chapter with regard to the technique of combination of mechanisms. A different matter is now under discussion.

It was stated that the bending moments M_i in the equation of virtual work, equation (3.3), could be any set in equilibrium with the external loads; they could, for example, be free bending moments as used in chapter 2. Suppose, then, that a trial collapse mechanism (δ_j, ϕ_i) is under examination, and it is required to write the collapse equation. If free bending moments have been tabulated, then an application of equation (3.3) gives

$$\Sigma W_j \delta_j = \Sigma (M_F)_i \phi_i, \tag{3.9}$$

where $(M_F)_i$ now represents a set of free bending moments in equilibrium with the applied loads W_j. The collapse equation for the mechanism may be written

$$\Sigma W_j \delta_j = \Sigma (M_p)_i |\phi_i|, \tag{3.10}$$

where $(M_p)_i$ is the numerical value of the full plastic moment at section i; for a non-uniform frame, the value of $(M_p)_i$ will of course be different at different sections. The terms $|\phi_i|$ in equation (3.10) indicate that the numerical values of the hinge rotations should be taken, and the sign ignored, since the work dissipated at a plastic hinge is always positive. Thus each term on the right-hand side of equation (3.10) is positive.

On combining equations (3.9) and (3.10),

$$\Sigma (M_F)_i \phi_i = \Sigma (M_p)_i |\phi_i|. \tag{3.11}$$

In this all-important equation, the external loading does not appear as such, but is represented by a set of free bending moments.

(The external loading can be represented by any one equilibrium set. It is of interest, but not of course of any real practical importance, that the *elastic* solution, being by definition an equilibrium set of moments, could be used on the left-hand side of equation (3.11). Thus the knowledge of the elastic solution to a frame leads at once to the plastic collapse equation. For the fixed-ended beam under eccentric point load, for example, figs. 2.2 and 3.2, the elastic solution is

$$\mathcal{M}_i \equiv \{\tfrac{2}{9}Wl, -\tfrac{8}{27}Wl, \tfrac{4}{9}Wl\},$$

while the collapse mechanism is

$$\phi_i \equiv \{\theta, -3\theta, 2\theta\}.$$

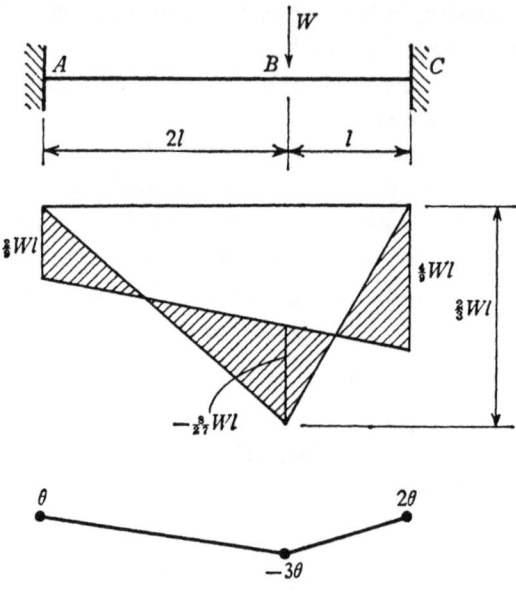

Fig. 3.2

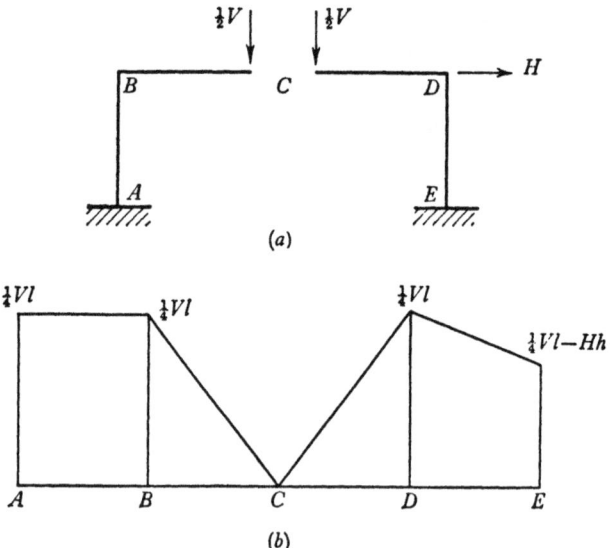

(a)

(b)

Fig. 3.3

Equation (3.11) thus gives

$$(\tfrac{2}{9}Wl)(\theta)+(-\tfrac{8}{27}Wl)(-3\theta)+(\tfrac{4}{9}Wl)(2\theta) = M_p(6\theta) \quad \text{or} \quad M_p = \tfrac{1}{3}Wl;$$

the result is, of course, evident from a direct inspection of fig. 3.2.)

For the fixed-base rectangular frame of span l and height h carrying idealized loads V and H as in the last chapter, the free bending moments of fig. 3.3 are obtained by cutting the frame at the centre of the beam (cf. figs. 2.47(a) and 2.48(a)). These free bending moments have values

$$(M_F)_i \equiv \{\tfrac{1}{4}Vl, \tfrac{1}{4}Vl, 0, \tfrac{1}{4}Vl, \tfrac{1}{4}Vl - Hh\}. \tag{3.12}$$

The mechanism of fig. 3.1(b) (statement (3.5) above), is

$$(\phi_i) \equiv \{\theta, 0, -2\theta, 2\theta, -\theta\}. \tag{3.13}$$

Thus the collapse equation corresponding to this mechanism is given from equation (3.11) as

$$(\tfrac{1}{4}Vl)(\theta)+(\tfrac{1}{4}Vl)(0)+(0)(-2\theta)+(\tfrac{1}{4}Vl)(2\theta)$$
$$+(\tfrac{1}{4}Vl-Hh)(-\theta) = M_p(6\theta), \tag{3.14}$$

or

$$\tfrac{1}{2}Vl+Hh = 6M_p, \tag{3.15}$$

as before.

Thus the equation of virtual work may be used to obtain direct plastic solutions from a table of free bending moments. There is no great advantage in this technique for the simple example just discussed, but it will be evident that considerable savings in computation will occur when more complex frames and loading patterns are examined. The examples of figs. 2.50 and 2.61 are reworked below.

There is, however, a further immediate application of the virtual work equation, which can again be illustrated by the example of the fixed-base rectangular frame subjected to the two point loads V and H. It was seen that there were three possible collapse modes (illustrated in fig. 2.43). To confirm the correctness of any one mode, it is necessary to check that the yield condition is satisfied everywhere in the frame. Thus if the combined mode of collapse of fig. 2.43(b), reproduced in fig. 3.1(b), had been assumed to be correct, then the collapse bending-moment distribution $(M_c)_i$ would be given by

$$(M_c)_i \equiv \{M_p, M_B, -M_p, M_p, -M_p\}. \tag{3.16}$$

It is necessary to check that the unknown bending-moment M_B is numerically less than M_p. The value of M_B must in fact be found by means of a statical analysis, and the equation of virtual work can be used for this analysis in the form

$$\Sigma(M_c)_i \psi_i = \Sigma(M_F)_i \psi_i. \qquad (3.17)$$

Here $(M_c)_i$ and $(M_F)_i$ are two sets of bending moments in equilibrium with the same external loads; $(M_c)_i$ happens to represent the (supposed) collapse state and $(M_F)_i$ a statically determinate state.

The hinge rotations (ψ_i) are, in accordance with the general principles of the virtual work equation, rotations of *any* proper mechanism. Thus (ψ_i) might be taken to correspond to the rotations of fig. 3.4,

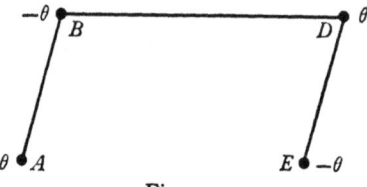

Fig. 3.4

$$(\psi_i) \equiv \{\theta, -\theta, 0, \theta, -\theta\}. \qquad (3.18)$$

Combining the two equilibrium states (3.12) and (3.16) with the hinge rotations of (3.18) by means of equation (3.17),

$$(M_p)(\theta) + (M_B)(-\theta) + (M_p)(\theta) + (-M_p)(-\theta)$$
$$= (\tfrac{1}{4}Vl)(\theta) + (\tfrac{1}{4}Vl)(-\theta) + (\tfrac{1}{4}Vl)(\theta) + (\tfrac{1}{4}Vl - Hh)(-\theta),$$

that is
$$M_B = 3M_p - Hh. \qquad (3.19)$$

On substituting the value of M_p from equation (3.15), the result of equation (2.48) is obtained:

$$M_B = \tfrac{1}{4}Vl - \tfrac{1}{2}Hh. \qquad (3.20)$$

Thus the virtual work equation may be used to determine not only the collapse equation of a frame, but also, by the consideration of various mechanisms, the complete statical analysis of the frame in that collapse state. These features will be illustrated with reference to the more complex design example of fig. 2.50, for which table 2.3 gives the free bending moments.

The order of working in the previous chapter will be followed. It was assumed first that the sidesway mode of fig. 3.4 might be correct. Thus, using equation (3.11), and cancelling θ throughout,

$$(96)(1) + (48)(-1) + (48)(1) + (0)(-1) = 4M_p,$$

or
$$M_p = 24, \qquad (3.21)$$

as was given in equations (2.52). A sketch of the net bending-moment diagram, fig. 2.52, now reveals that the solution is considerably in error, the largest bending moments occurring near the mid-section of the beam, and there is no point in making a complete statical analysis. As some measure of the error, the bending moment at the single point C will be computed. For this purpose, the mechanism of fig. 3.5 is suitable, since it involves only the three moments M_B, M_C and M_D, and for the assumed collapse mode of fig. 3.4, two are known, i.e. $M_B = -24$ and $M_D = 24$. Thus equation (3.17) gives

$$(-24)(1)+(M_C)(-2)+(24)(1) = (48)(1)+(0)(-2)+(48)(1),$$

or
$$M_C = -48. \qquad (3.22)$$

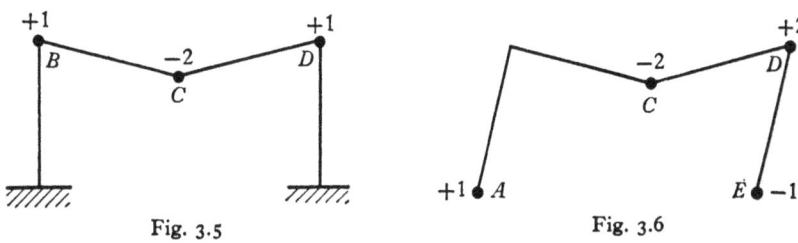

Fig. 3.5   Fig. 3.6

The mechanism of fig. 3.6 is now tried, still in conjunction with the table of free bending moments, table 2.3:

$$(96)(1)+(0)(-2)+(48)(2)+(0)(-1) = 6M_p,$$

or
$$M_p = 32. \qquad (3.23)$$

A sketch of the net bending-moment diagram reveals that this full plastic value is likely to be exceeded at the section B_3. To determine the value of this single bending moment, the mechanism of fig. 3.7 is convenient (the hinge rotations may be checked by computing the deflexion of B_3 from each of the two halves of the beam). However, the collapse mechanism of fig. 3.6 does

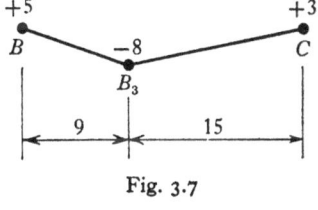

Fig. 3.7

not involve a hinge at B, so the moment there is not yet known. It can be found by using the virtual mechanism of fig. 3.5, which gives, from equation (3.17),

$$(M_B)(1)+(-32)(-2)+(32)(1) = (48)(1)+(0)(-2)+(48)(1),$$

or
$$M_B = 0. \qquad (3.24)$$

The mechanism of fig. 3.7 may now be used in equation (3.17):

$$(0)(5)+(M_{B_3})(-8)+(32)(3) = (48)(5)+(3)(-8)+(48)(3),$$

or
$$M_{B_3} = -33. \qquad (3.25)$$

Since the correct collapse mechanism has not been found, the new trial of fig. 3.8 is made:

$$(96)(5)+(3)(-8)+(48)(8)+(0)(-5) = 26M_p,$$

or
$$M_p = \tfrac{420}{13} = 32 \cdot 3. \qquad (3.26)$$

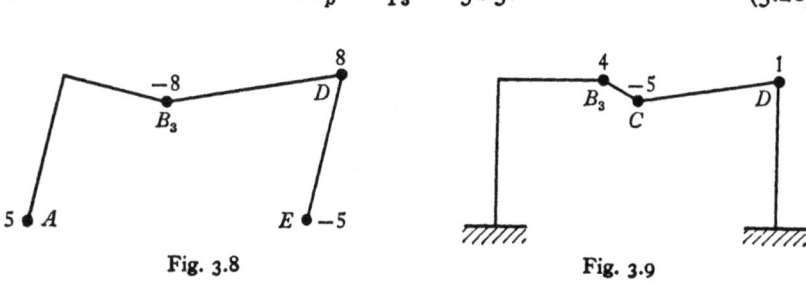

Fig. 3.8   Fig. 3.9

Virtual mechanisms of the type of fig. 3.9 may now be used to determine the bending moments throughout the frame; in this case, the value of the bending moment at C is given by

$$(-32 \cdot 3)(4)+(M_C)(-5)+(32 \cdot 3)(1) = (3)(4)+(0)(-5)+(48)(1),$$

or
$$M_C = -31 \cdot 4. \qquad (3.27)$$

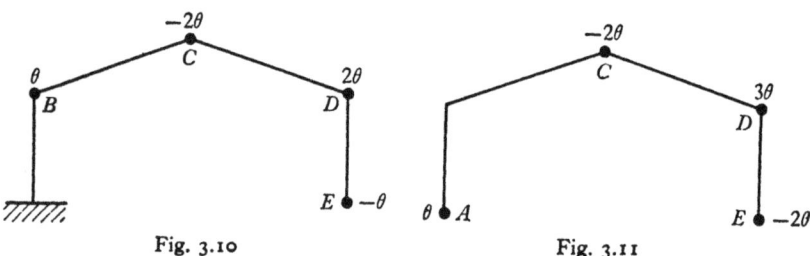

Fig. 3.10   Fig. 3.11

Exactly the same techniques may be used for the pitched-roof frame, fig. 2.61, but there are some geometrical difficulties. The hinge rotations of fig. 3.10 may be written down from the general expressions given in fig. 2.58, but those of fig. 3.11 are not quite so easily determined by a direct method. Similarly, it is not easy to see at a glance that the hinge rotations of fig. 3.12 are correct; an easy way of computing such rotations is given in the next chapter.

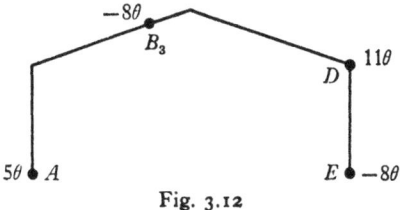

Fig. 3.12

The mechanism of fig. 3.10, applied to the table of free moments, table 2.3, gives, as the corresponding collapse equation,

$$(48)(1)+(0)(-2)+(48)(2)+(0)(-1) = 6M_p,$$

or
$$M_p = 24, \tag{3.28}$$

as in equation (2.68). Similarly, the mechanism of fig. 3.11 gives

$$(96)(1)+(0)(-2)+(48)(3)+(0)(-2) = 8M_p,$$

or
$$M_p = 30, \tag{3.29}$$

as in equation (2.70). Finally, the correct mechanism of fig. 3.12 gives

$$(96)(5)+(3)(-8)+(48)(11)+(0)(-8) = 32M_p,$$

or
$$M_p = 30{\cdot}75. \tag{3.30}$$

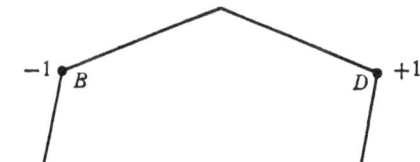

Fig. 3.13

As before, the complete bending-moment diagram for this correct collapse state may be constructed by considering virtual mechanisms of the type shown in fig. 3.13. This particular mechanism gives

$$(30{\cdot}75)(1)+(M_B)(-1)+(30{\cdot}75)(1)+(-30{\cdot}75)(-1)$$
$$= (96)(1)+(48)(-1)+(48)(1)+(0)(-1),$$

or $$M_B = -3\cdot75,\qquad(3.31)$$

while the mechanism of fig. 3.14 gives

$$(-3\cdot75)(1)+(-30\cdot75)(-4)+(M_C)(3)$$
$$= (48)(1)+(3)(-4)+(0)(-3),$$

or $$M_C = -27\cdot75.\qquad(3.32)$$

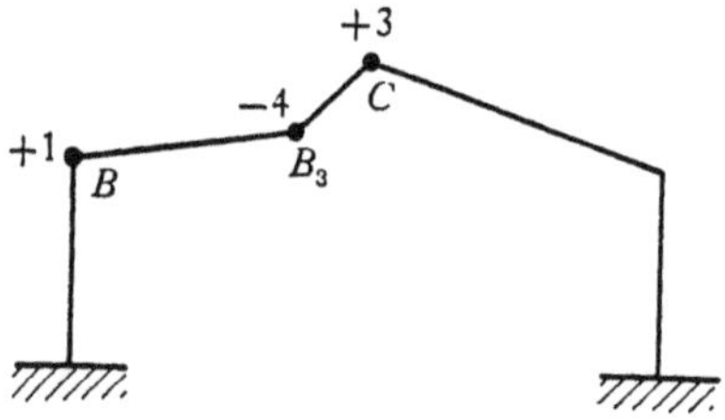

Fig. 3.14

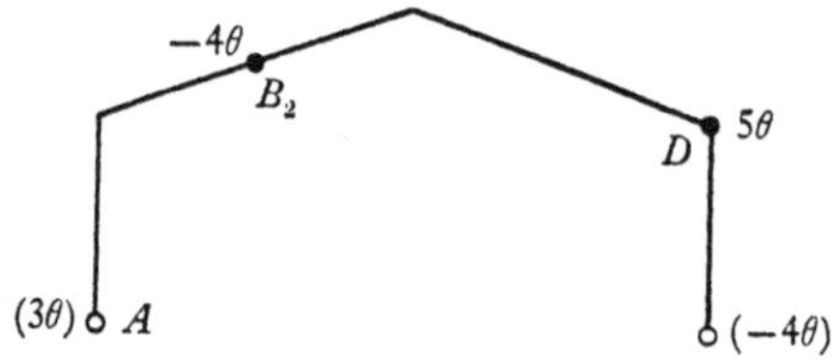

Fig. 3.15

Finally, collapse of the pin-based frame carrying the same loading can also be checked quickly by using the equation of virtual work. The correct collapse mechanism is shown in fig. 3.15, and this gives, again in conjunction with the table of free moments, table 2.3,

$$(96)(3)+(12)(-4)+(48)(5)+(0)(-4) = 9M_p,$$

or $$M_p = 53\cdot3.\qquad(3.33)$$

In this equation, notice that there are contributions from the pinned feet to the *virtual* work on the left-hand side, but no contributions to the *real* work on the right-hand side, since the actual bending moments are zero at the feet.

The virtual work equation is an enormous aid to calculation once the correct collapse mechanism is found, but finding the right mechanism by trial and error is often not an easy task. The next chapter deals with this problem, making use of the basic theorems of plastic design, which will now be discussed.

3.2   The fundamental theorems

The basic theorems of plastic theory are 'obvious' enough, and have really been tacitly assumed in all the foregoing work. Formal proofs will now be given, both because a firm foundation will thereby be provided to the theory, and also because the theorems immediately indicate ways in which further analytical techniques can be developed.

For the purposes of the present volume, what is known as *proportional loading* only will be considered. The loads acting on a structure will not be allowed to vary randomly and independently, but, on the contrary, can all be specified in terms of one of their number. That is, each load might be thought of as having its *working* value, that value being multiplied, as the load increases, by a common factor λ, the load factor. The fundamental theorems to be stated here are concerned with the value λ_c of the load factor at collapse of the structure.

The first 'obvious' theorem is that λ_c has a definite value; as the loads on the structure are slowly increased, that is, as the value of λ is increased, collapse occurs at one certain value λ_c. This is the uniqueness theorem.

The second theorem is concerned with the value of λ resulting from the analysis of a guessed mechanism of collapse. If the guess happens to be correct, then, of course, $\lambda = \lambda_c$; otherwise, the theorem states that the value of λ will always be greater than the true value λ_c, or at best equal to λ_c. Again, imagining all the loads to be increased slowly in proportion from their working values, collapse will occur by the guessed mechanism unless it has *already* occurred by the correct mechanism. This is the 'unsafe theorem'.

The third theorem is concerned with equilibrium states of the structure. Suppose that it is possible to find at a certain load factor λ a bending-moment distribution for the frame such that the full plastic moment is not exceeded at any cross-section. The bending-moment distribution need not be the actual distribution at the load factor λ; it must only be in *equilibrium* with the external loads at that load factor. That is, it is a system of free bending moments combined with any proper set of reactant moments. Such a bending-moment distribution will not, in general, correspond to a mechanism, so that collapse is not implied; at worst, if the bending-moment distribution happens to be the collapse distribution, then collapse is just occurring. The theorem states that the value of λ will be less than λ_c, or at most equal to λ_c. That is, if it is *possible* to find a bending-moment distribution which does not correspond to a collapse mechanism, then the structure *will* not collapse at that load factor by any mechanism. This is the 'safe theorem', and it is of particular use in design rather than in analysis.

In fact, both the 'unsafe' and the 'safe' theorems, stated here in terms of load factor so that they correspond to the problem of analysis of a given structure, can be 'inverted' to correspond to the problem of design. The basic collapse equation relates loads and full plastic moments, and is of the form $\lambda_c Wl = kM_p$, where k is a numerical constant. The analytical problem is to find, for a given frame with known values of M_p, the value of the collapse load factor. The design problem is to assign sections to the various members of the frame so that the working loads can be safely carried at a specified load factor.

Thus, interpreting the theorems in a design sense, the unsafe theorem states that the value of M_p resulting from the analysis of an arbitrarily assumed mechanism of collapse will be smaller than that actually required. Similarly, the use of the safe theorem will lead to an estimate of M_p that is larger than the value required.

Collapse requirements. It was noted in chapter 2 (statements (2.7)) that a frame at collapse had to satisfy three conditions. First, there must be formed a sufficient number of plastic hinges to turn the frame, or part of it, into a mechanism; this was called the *mechanism condition*.

Second, the frame is at all stages in equilibrium. The bending-moment distribution must be such as to satisfy equilibrium with the external loads; this was called the *equilibrium condition*.

Finally, it has been seen that, for a perfectly plastic material, there is a limiting value to the bending moment which can be attained at any cross-section, called the full plastic moment. At collapse of a frame, the bending-moment distribution must be such that the value of the bending moment at any section does not exceed the full plastic value at that section; this was called the *yield condition*.

If these three conditions are satisfied simultaneously, then the corresponding value of the load factor is unique.

Uniqueness theorem. To show that there is a single value only, λ_c, of the collapse load factor, it will be assumed in the first instance that this is not the case. Suppose then that it were possible to find two different collapse mechanisms for a particular frame under given loading, formed at different load factors λ^* and λ^{**}. For the first of these collapse mechanisms, for which joint displacements δ_j^* are compatible with hinge rotations θ_i^*, suppose that the collapse bending-moment distribution is denoted by the set of bending moments M_i^*. These bending moments are in equilibrium with external loads $\lambda^* W_j$ and, moreover, they satisfy the yield condition. Thus the mechanism condition is satisfied by postulating the mechanism (δ^*, θ^*)—the suffixes i and j have been dropped—the equilibrium condition by writing the corresponding equilibrium set $(\lambda^* W, M^*)$, and the yield condition by the statement that $|M^*| \leqslant M_p$. Using a similar notation for the second mechanism, the following statements may be made:

$$A: (\lambda^*W, M^*) \text{ is an equilibrium set;}$$
$$B: (\lambda^{**}W, M^{**}) \text{ is an equilibrium set;}$$
$$C: (\delta^*, \theta^*) \text{ is a compatible mechanism;}$$
$$D: (\delta^{**}, \theta^{**}) \text{ is a compatible mechanism.}$$

(3.34)

The collapse equation for the first mechanism may be written by combining statements A and C in the equation of virtual work:

$$\Sigma\lambda^*W\delta^* = \Sigma M^*\theta^*.$$ (3.35)

On the left-hand side of this equation, the factor λ^* is common to all the loading terms (proportional loading) and can be taken outside the summation sign. On the right-hand side the value of M^* is known at each hinge position to be equal to $\pm M_p$ at that cross-section. Since the plastic work dissipated at each hinge is positive (that is, the sign of the bending moment corresponds to that of the hinge rotation), each term on the right-hand side of equation (3.35) is positive, and the equation may be written

$$\lambda^*\Sigma W\delta^* = \Sigma M_p|\theta^*|.$$ (3.36)

Statements B and C in (3.34) may also be combined by using the equation of virtual work:

$$\Sigma\lambda^{**}W\delta^* = \Sigma M^{**}\theta^*.$$ (3.37)

On the left-hand side, the common factor λ^{**} can again be taken outside the summation sign. On the right-hand side, the bending moments M^{**} satisfy the yield condition; that is, if the two mechanisms θ^* and θ^{**} have a common hinge at a certain cross-section i, then M_i^{**} will equal $\pm(M_p)_i$, but otherwise $|M_i^{**}| < (M_p)_i$. Thus each of the terms $M_i^{**}\theta_i^*$ on the right-hand side of equation (3.37) will be less than, or at most equal to, $(M_p)_i|\theta_i^*|$, so that the equation leads to

$$\lambda^{**}\Sigma W\delta^* \leqslant \Sigma M_p|\theta^*|.$$ (3.38)

Comparison of equation (3.36) with inequality (3.38) shows that

$$\lambda^{**} \leqslant \lambda^*.$$ (3.39)

Of the four statements (3.34), the last, D, describing the second mechanism of collapse, has not yet been used. By combining B with D and A with D, exactly as above, it is seen that

$$\lambda^* \leqslant \lambda^{**}.$$ (3.40)

Inequalities (3.39) and (3.40) can be satisfied simultaneously only if the two load factors have the same value; this value is λ_c, that of the unique collapse load factor.

It has not been established that the collapse mechanism is unique, or that the bending-moment distribution at collapse is unique. Indeed, it is possible that alternative mechanisms of collapse can co-exist. If they do so, however, then they must be formed at the same load factor λ_c.

The unsafe theorem. If the collapse equation is written for an arbitrarily assumed mechanism, then the resulting value of the load factor, λ', is always greater than, or at best equal to, the true load factor λ_c. The following statements will be used:

$$A: (\lambda_c W, M_c) \text{ is the actual collapse distribution;}$$
$$B: (\delta, \theta) \text{ is the assumed collapse mechanism.} \tag{3.41}$$

The collapse equation written for the assumed collapse mechanism is

$$\lambda' \Sigma W \delta = \Sigma M_p |\theta|; \tag{3.42}$$

on the right-hand side, it should be noted that the value of M_p has been taken at every hinge position. Statements A and B of (3.41), when combined by the equation of virtual work, give

$$\lambda_c \Sigma W \delta = \Sigma M_c \theta. \tag{3.43}$$

Now the values of M_c satisfy, by definition, the yield condition; that is, $|M_c| \leqslant M_p$. Thus equation (3.43) leads to the inequality

$$\lambda_c \Sigma W \delta \leqslant \Sigma M_p |\theta|, \tag{3.44}$$

and, comparing (3.42) and (3.44),

$$\lambda' \geqslant \lambda_c. \tag{3.45}$$

The safe theorem. If a set of bending moments can be found at a load factor λ'' to satisfy both the equilibrium and yield conditions, then λ'' is always less than, or at most equal to, the true load factor λ_c. The following statements will be used:

$A: (\lambda'' W, M'')$ represents a set of bending moments, in equilibrium with the applied loads, which satisfies the yield condition;

$B: (\lambda_c W, M_c)$ is the actual collapse distribution;

$C: (\delta_c, \theta_c)$ is the actual collapse mechanism.

$$\tag{3.46}$$

The value of λ_c is, of course, given by

$$\lambda_c \Sigma W \delta_c = \Sigma M_c \theta_c = \Sigma M_p |\theta_c|. \tag{3.47}$$

If statements A and C of (3.46) are combined,

$$\lambda'' \Sigma W \delta_c = \Sigma M'' \theta_c, \tag{3.48}$$

and, since $|M''| \leqslant M_p$, comparison of (3.47) and (3.48) shows that

$$\lambda'' \leqslant \lambda_c. \tag{3.49}$$

The three theorems can be displayed as follows:

$$\lambda = \lambda_c \begin{cases} \text{MECHANISM CONDITION} & \lambda \geqslant \lambda_c, \\ \text{EQUILIBRIUM CONDITION} \\ \text{YIELD CONDITION} \end{cases} \left. \begin{array}{l} \\ \\ \end{array} \right\} \ \lambda \leqslant \lambda_c. \tag{3.50}$$

It should be noted that equilibrium need not be satisfied in deriving an unsafe solution from an assumed mechanism; the mechanism could, for example, have extra hinges for which no possible equilibrium distribution of bending moments could be constructed. Mechanisms of this type are mentioned in the next chapter; see fig. 4.4.

Before applying the theorems to some numerical examples, one or two corollaries may be noted. As a first result of the theorems, the definite statement can be made that if material (of negligible self-weight) is added to a frame, or a restraint imposed, the frame cannot thereby be weakened. This 'obvious' corollary is not so self-evident to a designer accustomed to notions of elastic stresses; the increase of section of a member may well increase rather than decrease the stress acting in that member.

The proof of the corollary follows quickly from the theorems. Since the bending-moment distribution for the collapse state of the unstrengthened frame must satisfy the yield condition, the same bending-moment distribution must certainly satisfy the yield condition for the strengthened frame. The safe theorem then states that the strengthened frame cannot collapse at a load less than that of the unstrengthened frame. (Similarly, the removal of material from a frame, or the removal of a constraint, can only weaken the frame; this may be seen by consideration of the unsafe theorem.)

The safe theorem is, whether he knows it or not, the fundamental tool of the conventional designer. For example, it is usual to assume in the conventional elastic design of steel frames that each floor beam is simply supported, and can be designed in isolation from its neighbours, even though it is, in fact, fairly rigidly connected at its ends by bolts and cleats. If the beam can be demonstrated to be satisfactory in the pin-ended state, then it cannot be weakened if, in practice, the ends are bolted up so as to transmit moment.

A great deal of design is based upon plausible equilibrium states of the structure, either derived crudely, as in the case of the pin-ended beam just considered, or by more sophisticated methods. If an equilibrium state is used to proportion the members so that the yield condition is satisfied, then a safe design results. It is this fact which justifies the use in engineering design of simplifying assumptions, and it is an explanation of the fact that buildings designed in accordance with 'irrational' codes of practice are usually safe structures.

An important corollary of the unsafe theorem is that the true load factor at collapse is the smallest possible that can be found from a consideration of all possible mechanisms of collapse. This is an important consideration in the method of combination of mechanisms, described in the next chapter.

3.3   Upper and lower bounds

The safe theorem and the unsafe theorem may be used together to furnish bounds on the value of the load factor, if the problem is one of analysis, or on the value of M_p, if the problem is one of design. As a first example, the design of the rectangular portal frame, fig. 2.50, will be considered. The first trial mechanism, fig. 2.43(a), gave a value of M_p equal to 24; by the unsafe theorem, the true value of M_p must be larger than this (or at best equal to 24, had the sidesway mechanism been correct, which was not in fact the case). The complete bending-moment diagram recorded in table 2.4 (fig. 2.52), shows that the yield condition is violated at several cross-sections, the worst violation occurring at section B_3, where the moment has value 51. Now the bending moments in table 2.4 are certainly in equilibrium with the applied loads, and, if M_p is set equal to 51, they will not violate the yield condition (but the mechanism condition will no longer be satisfied). By the safe theorem, the true value of M_p must be less than 51. The table of bending moments, table 2.4, in fact leads immediately to

$$24 \leqslant M_p \leqslant 51. \tag{3.51}$$

These bounds are too wide for practical design purposes.

The second trial mechanism, fig. 2.43(b), leading to the bending moments of table 2.5, furnish the much closer bounds

$$32 \leqslant M_p \leqslant 33, \tag{3.52}$$

and these are certainly close enough for practical design.

Similar bounds may be found for each stage of the working of the example of the pitched-roof frame, fig. 2.61.

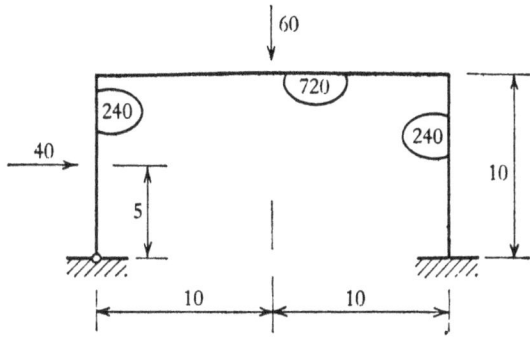

Fig. 3.16

As an example of the analysis rather than design of a frame, the unsymmetrical portal of figure 3.16 will be discussed. The frame has unequal beam and columns, and carries the working loads shown; the value of the load factor at collapse is required. A start may be made on this problem by considering some simple mechanisms of collapse. For example, the sidesway mode of fig. 3.17 gives

$$(40\lambda)(5\theta) = (240)(3\theta),$$

or $$\lambda = 3{\cdot}6. \tag{3.53}$$

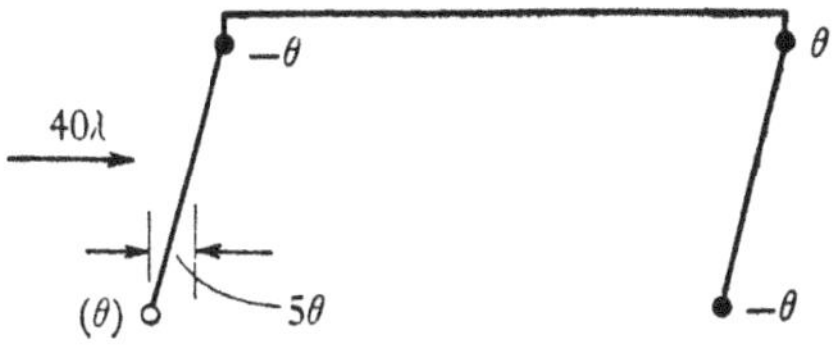

Fig. 3.17

137

The local collapse mode of fig. 3.18 gives

$$(40\lambda)(5\theta) = (240)(3\theta),$$

or
$$\lambda = 3 \cdot 6, \qquad (3 \cdot 54)$$

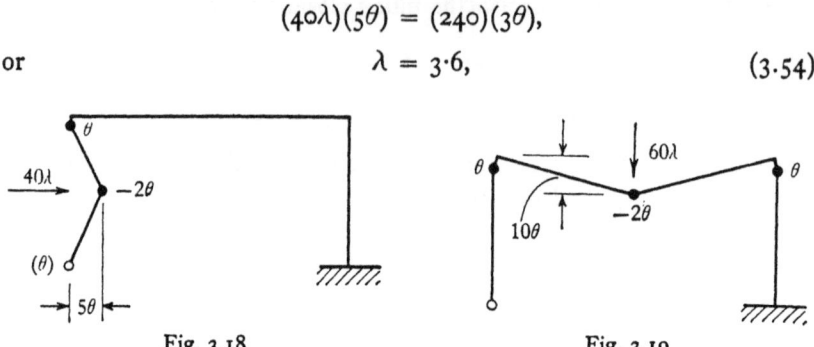

Fig. 3.18   Fig. 3.19

and the local collapse mode of fig. 3.19 gives

$$(60\lambda)(10\theta) = (720)(2\theta) + (240)(2\theta),$$

or
$$\lambda = 3 \cdot 2. \qquad (3 \cdot 55)$$

So far, then, it has been established that $\lambda_c \leqslant 3 \cdot 2$; the mode of fig. 3.19 will now be analysed to determine whether or not the yield condition is satisfied. Fig. 3.20 displays the bending moments in the frame at a load

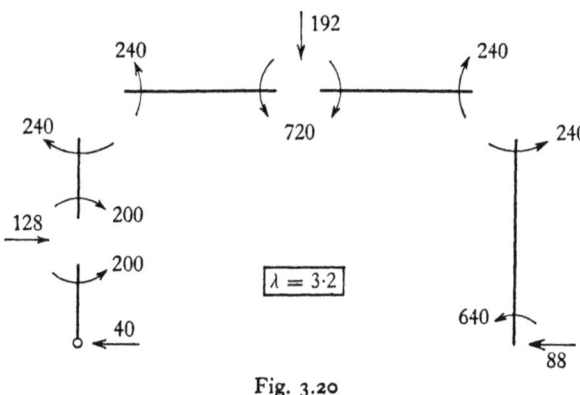

Fig. 3.20

factor of $3 \cdot 2$, with hinges occurring as in fig. 3.19. The bending moments in the beam can therefore be entered immediately; by taking moments for the left-hand column about its top, the shear force (40) can be determined at the foot of the column. Thus the shear force at the foot of the other column must be 88, since the total factored horizontal load on the frame is 128, and the bending moments can be computed in both columns. It will be seen that the yield condition is violated at the foot of the right-hand column, so that the assumed mechanism of collapse is not correct.

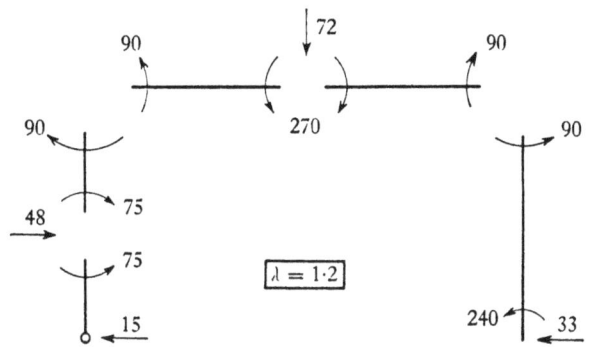

Fig. 3.21

Suppose that every number in fig. 3.20 is multiplied by $240/640 = 3/8$. The result is shown in fig. 3.21, and it will be seen that an equilibrium bending-moment distribution has been derived, at a load factor of 1·2, which just satisfies the yield condition. Figure 3.21 therefore represents a safe solution, and hence

$$1\cdot2 \leqslant \lambda_c \leqslant 3\cdot2. \tag{3.56}$$

The reader may wish to check that a similar analysis of the collapse mechanism of fig. 3.17 leads to the bounds $1\cdot8 \leqslant \lambda_c \leqslant 3\cdot6$.

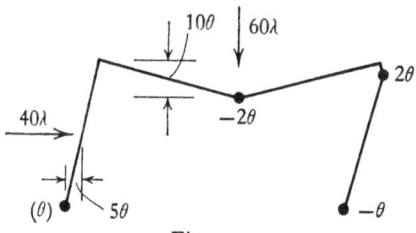

Fig. 3.22

The bending-moment distribution of fig. 3.20 indicates that the mechanism of fig. 3.22 might give a better estimate of the collapse load factor. The collapse equation is

$$(40\lambda)(5\theta) + (60\lambda)(10\theta) = (720)(2\theta) + (240)(3\theta),$$

or $$\lambda = 2\cdot7. \tag{3.57}$$

However, a statical analysis, fig. 3.23, shows that the yield condition is violated in the left-hand column. A safe solution may be obtained by scaling fig. 3.23 in the ratio $240/300 = 0.8$, so that

$$2.16 \leqslant \lambda_c \leqslant 2.7. \qquad (3.58)$$

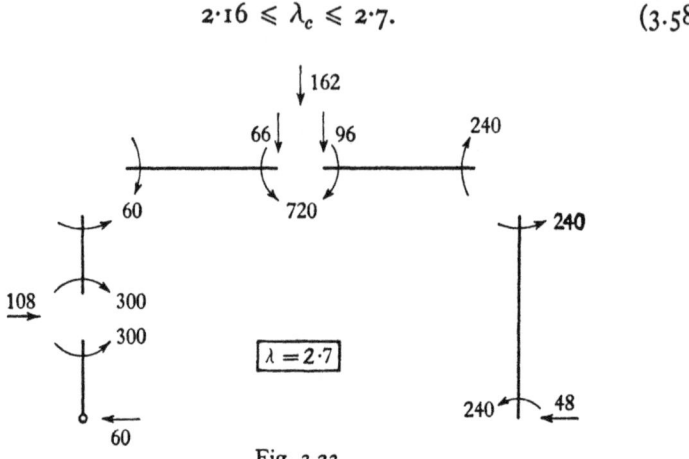

Fig. 3.23

The correct mechanism of collapse is sketched in fig. 3.24. The collapse equation is

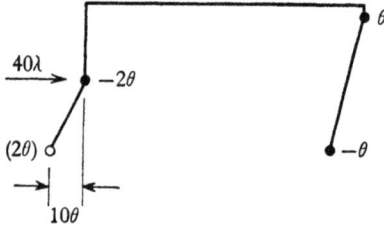

$$(40\lambda)(10\theta) = (240)(4\theta),$$

or $\qquad \lambda = 2.4, \qquad (3.59)$

Fig. 3.24

and the statical analysis displayed in fig. 3.25 confirms that the yield condition is not violated at any cross-section.

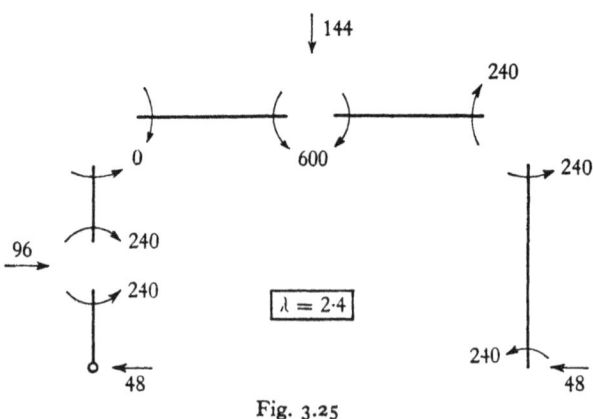

Fig. 3.25

lt will have been appreciated that even this relatively simple problem has presented some difficulties when tackled by a trial-and-error method, although bounds on the correct answer were found at each stage. For more complex frames where the number of possible collapse mechanisms is correspondingly very much larger, a more orderly method is needed. Such a method is given in the next chapter.

EXAMPLES

3.1 What are the maximum and minimum principles as applied to ductile frame structures in the plastic range? Discuss their implications, and their use in the analysis and design of structures. Illustrate your answer by reference to the rigid-jointed frame shown. (*M.S.T.* II, 1955.)

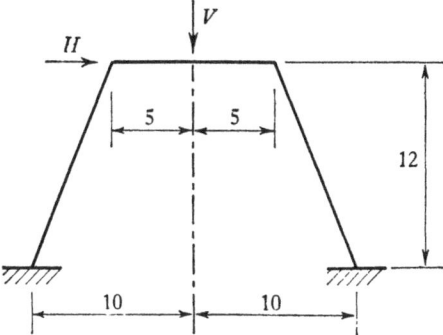

3.2 The rigid frame shown is of constant section, has fixed feet and supports a uniformly distributed load $2W$ and a horizontal point load W.

Illustrate diagrammatically all the possible collapse mechanisms for this frame.

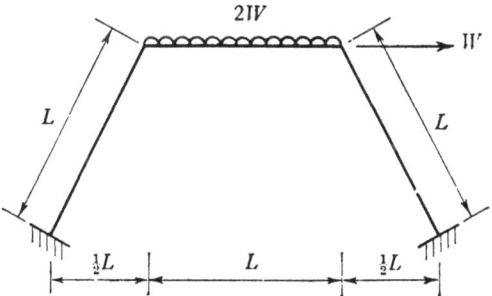

Calculate the required plastic moment of resistance M_p for the frame section, or alternatively upper and lower bounds to its value in terms of L and the loading W. (University of London, B.Sc. (Engineering) Part III: Civil and Municipal, University College, 1962.)   (*Ans.* $M_p = 0.192WL$.)

3. THE TOOLS OF PLASTIC DESIGN

3.3 The uniform frame shown will collapse by one of its three basic modes. Considering collapse to occur by pure sidesway, a value of $W(W_1)$ can be determined from the principle of virtual work, and this value will be an upper bound on the actual collapse load of the frame. For this mechanism, and using the derived value of W_1, a statical analysis may be made, from which it will be

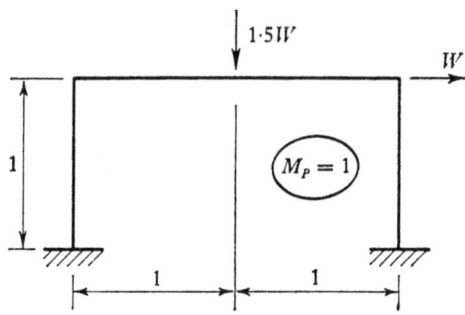

found that the moment at the centre of the beam is kM_p, where $k > 1$. If all the moments in the frame are scaled down in the ratio $1/k$, a bending-moment distribution is obtained that is an equilibrium distribution for loads W_1/k, and which nowhere exceeds the fully plastic moment ('statically admissible' distribution). Hence W_1/k is a lower bound on the collapse load. Repeat the analysis, assuming collapse to occur (a) in the beam alone, and (b) by the combined mechanism.

3.4 Find bounds on the value of W to cause collapse of the frame shown for (a) $P = \frac{1}{4}W$, and (b) $P = W$.

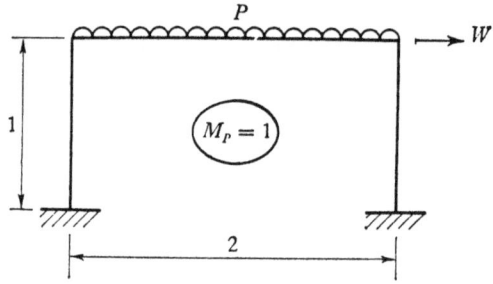

3.5 The square truss shown is composed of six bars all of the same section. Each bar can carry a maximum load of P_0 in either tension or compression, and can extend indefinitely under this constant load. The joints of the truss can transmit forces but not bending moments. By considering points A, B and C to be fixed, derive an upper bound on the collapse value of P, assuming a mechanism to be formed by yielding in bars AD, BD and CD. Hence determine a lower bound on the collapse load. What is the actual collapse mechanism and the actual collapse load?

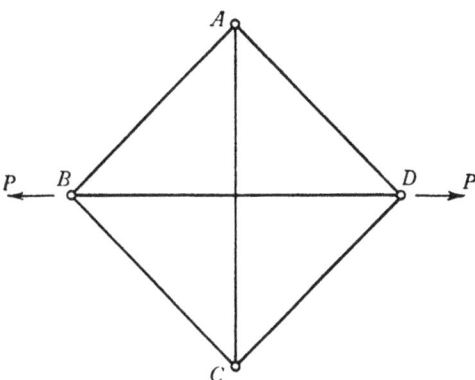

3.6 If one of the bars is removed from the truss in example 3.5, the truss becomes statically determinate and the bar forces may be found. Such a system of bar forces is an equilibrium system for the original truss, and will be statically admissible if none of the bar forces exceeds P_0. Remove the bars *AB*, *AC* and *BD* in turn and find lower bounds.

3.7 By considering various arrangements of seven hinges for the frame shown, determine upper and lower bounds on the factor λ by which the loads must be multiplied in order that collapse will just occur. The fully plastic moments of the various members are given in the figure.

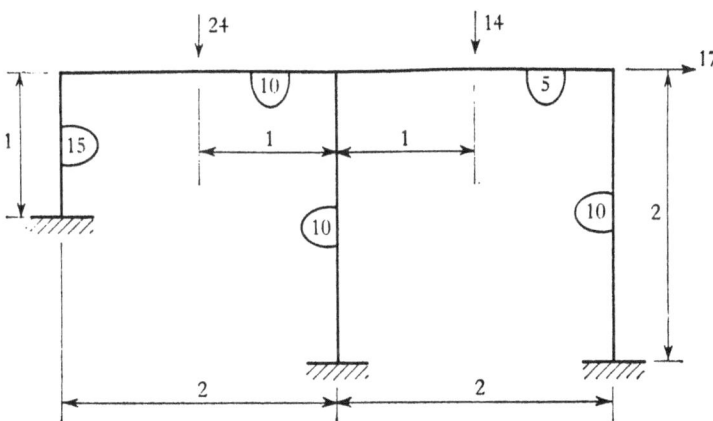

4

THE COMBINATION OF MECHANISMS

4.1 Basis of the method

It was seen in chapter 3 that the equation of virtual work could be used
to derive an equilibrium equation for a frame. Figures 2.42 and 2.43
are redrawn in figs 4.1 and 4.2, and the virtual work equation applied
to the mechanism of fig. 4.2(a) leads at once to the relationship

$$Hh = M_A - M_B + M_D - M_E. \qquad (4.1)$$

(This equation also follows from a consideration of the shear equilibrium of the whole frame.)

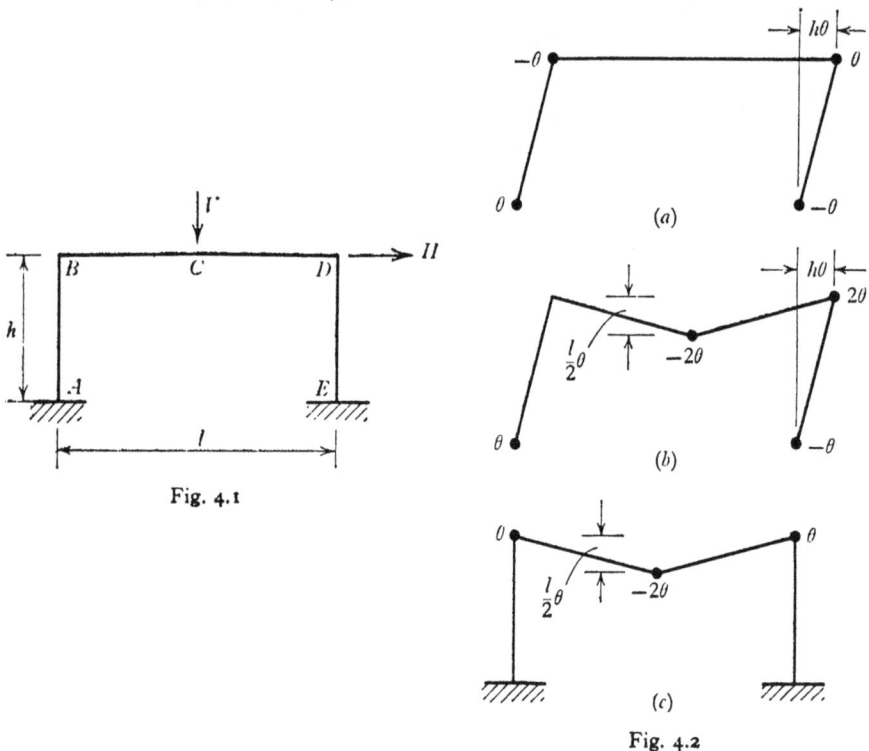

Fig. 4.1

Fig. 4.2

Equation (4.1) must be satisfied whatever the state of the frame,
elastic or plastic. If M_A to M_E are the values of the bending moments

at the sections A, B, C, D and E, they are always related by equation (4.1). The question then arises as to how many such equilibrium relationships hold for any given frame.

In fig. 4.1 the five sections A to E are *critical sections* at which plastic hinges might form under the particular loading system. For the type of frame considered here, critical sections will occur at loading points and at joints, since the bending-moment diagram will consist of straight lines between these cross-sections. Thus if the bending moments were known at the five sections A to E, the complete bending-moment diagram could be constructed. More generally, therefore, the five sections may be regarded as *cardinal sections*; the complete state of the frame at any stage may be described in terms of the values of the bending moments at the cardinal sections.

The frame has three redundancies and, looked at from another point of view, the bending-moment diagram would be completely specified if the values of those three redundancies were known. Now the redundancies could quite properly be chosen as three of the five unknown bending moments M_A to M_E; if they were so chosen, then the conclusion must be that the other two bending moments must be calculable.

From equation (4.1), for example, if the values of M_A, M_B and M_D were known, then the value of M_E could be calculated immediately. Further, there must exist some other equilibrium relationship which would enable the value of M_C to be calculated. This second equilibrium equation can be written by using the mechanism of fig. 4.2(c) in the virtual work equation:

$$\tfrac{1}{2}Vl = M_B - 2M_C + M_D. \tag{4.2}$$

(This equation also follows from a consideration of the equilibrium of beam BCD.)

The third mechanism, fig. 4.2(b), will also lead to an equation of equilibrium:
$$Hh + \tfrac{1}{2}Vl = M_A - 2M_C + 2M_D - M_E. \tag{4.3}$$

However, it will be seen that equation (4.3) contains no new information; it is simply the result of adding together equations (4.1) and (4.2). Only two of these three equations are independent; if equations (4.1) and (4.2) are thought of as being 'basic', then equation (4.3) can be derived by addition, or if equations (4.1) and (4.3) are regarded as the independent equations, then equation (4.2) can be derived by subtraction.

From this simple example can be deduced the general rule governing the number of independent equilibrium equations for a frame. If the

4. THE COMBINATION OF MECHANISMS

frame has a number N of critical (cardinal) sections, and has a number R of redundancies, then there will exist $(N-R)$ independent relationships between the values of the bending moments at the cardinal sections. All other equations relating the values of the bending moments can be deduced from these $(N-R)$ independent equilibrium equations.

Now an equilibrium equation such as (4.1) was derived directly from a mechanism (fig. 4.2(a)); there is, in fact, an exact correspondence between an equilibrium equation and a mechanism of collapse, and the previous paragraph can be 'translated' as follows:

If a frame has a number N of critical sections, and has a number R of redundancies, then there will exist $(N-R)$ independent mechanisms of collapse. All other mechanisms of collapse can be deduced from these $(N-R)$ independent mechanisms.

From this statement stems the method of combination of mechanisms for the plastic analysis of frames.

It was seen that equations (4.1) and (4.2) summed to equation (4.3); this summation can be seen also by an examination of the mechanisms sketched in fig. 4.2. The hinge rotations of fig. 4.2(a) and (c) sum immediately to those of fig. 4.2(b). The *cancellation* of the hinge at the top of the windward column, B, is not accidental; indeed, it is of prime importance, as will be seen.

Equations (4.1) to (4.3) have been written in general terms, but they can, of course, be simplified for the examination of plastic collapse. Regarding the mechanisms of fig. 4.2 not as virtual mechanisms, but as plastic collapse mechanisms, the corresponding collapse equations for fig. 4.2(a) and (c) are

$$(a) \quad Hh = 4M_p, \qquad (c) \quad \tfrac{1}{2}Vl = 4M_p. \qquad (4.4)$$

When these two equations are added together, to give the collapse equation for the mechanism of fig. 4.2(b),

$$(b) \quad Hh + \tfrac{1}{2}Vl = 6M_p, \qquad (4.5)$$

an adjustment must be made on the right-hand side, the total of $8M_p$ being reduced to $6M_p$. This reduction is associated with the cancellation of hinge B when the mechanisms are combined.

Equation (4.4(a)), for example, may be written in full as

$$Hh\theta = M_p\theta + M_p\theta + M_p\theta + M_p\theta, \qquad (4.6)$$

each of the contributions from the hinges at A, B, D and E being shown separately on the right-hand side. As usual, all the terms are positive. Similarly, there is a contribution of $M_p\theta$ from hinge B to the second of equations (4.4), and this is again positive. Since hinge B disappears in the combined mechanism, fig. 4.2(b), a deduction of $M_p\theta$ must be made from *each* of the two equations that are being summed. In general, when two mechanisms are combined and a hinge cancelled, due care must be taken to remove the plastic work terms corresponding to the cancelled hinge. The numerical adjustment is simple, and will perhaps be grasped most easily by means of examples.

The frame of fig. 4.3 has uniform section ($M_p = 15$) and carries the loads shown; the value of the load factor λ is required at collapse. To repeat the numerical work establishing the number of independent mechanisms, there are five critical sections and three redundancies, so that there must be two independent mechanisms, which will be taken as those of fig. 4.2(a) and (c). The two independent collapse equations are therefore

$$
\begin{array}{ll}
(a) & 20\lambda = 60; \quad \lambda = 3, \\
(c) & 30\lambda = 60; \quad \lambda = 2.
\end{array}
\quad\quad (4.7)
$$

and

The conclusion may be drawn that the sidesway mechanism will not occur, since it forms at a load factor of 3, whereas the beam collapses at a load factor of 2. Since the correct answer to the problem will be obtained for the mechanism which gives the smallest value of λ, combinations of the independent mechanisms must be examined to see if these give a value of less than 2. It was seen that only one combination of the two mechanisms is possible, involving the cancellation of the hinge at B; i.e. there are only the three mechanisms of collapse shown in fig. 4.2 for the rectangular portal frame under this type of loading. ('Ridiculous' mechanisms have been excluded. It is not always possible to state *a priori* in which way a frame will sway, but for the loading of fig. 4.3 it is clear that a sway from right to left can be legitimately excluded from consideration. Similarly, the centre C of the beam will not move vertically *upwards*, although the mechanism of fig. 4.2(c) could in theory be reversed.)

The calculations leading to the collapse equation for mode (b) of fig. 4.2 can be laid out conveniently as follows:

$$
\begin{array}{ll}
(a) & 20\lambda = 4(15); \quad \lambda = 3{\cdot}0, \\
(c) & 30\lambda = 4(15); \quad \lambda = 2{\cdot}0, \\
\hline
& 50\lambda \quad 8(15) \\
\end{array}
\quad\quad (4.8)
$$

Cancel hinge B:

$$
\begin{array}{ll}
& 2(15) \\
\hline
(b) & 50\lambda = 6(15); \quad \lambda = 1{\cdot}8.
\end{array}
$$

4. THE COMBINATION OF MECHANISMS

For the particular numerical values of this example, therefore, the mechanism of fig. 4.2(b) gives the correct collapse mode.

It should be noted that the cancellation of hinges is the key to the method of combination of mechanisms. First, if two equations such as (4.7) are added in any proportions without any reduction of the right-hand side, a value of λ will result which is intermediate between the two original values. Thus the mechanisms cannot possibly be combined to give a smaller value of λ unless some terms are cancelled.

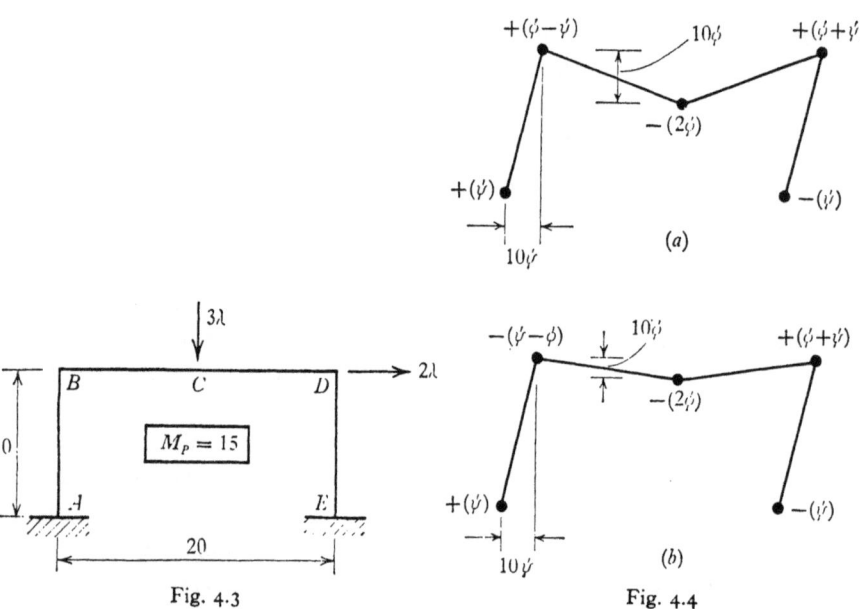

Fig. 4.3   Fig. 4.4

Second, if two mechanisms, each of one degree of freedom, are superimposed in arbitrary proportions, a mechanism of two degrees of freedom will result. For example, if the mechanisms of fig. 4.2(a) and (c) are supposed to have different hinge rotations ψ and ϕ respectively (instead of the common value θ), then the mechanisms of Fig. 4.4 will be obtained, depending on whether $\phi \lessgtr \psi$. It is reasonable to suppose that both ϕ and ψ are positive (which would exclude the 'ridiculous' type of mechanism), in which case fig. 4.4 gives the most general pattern of deformation that is possible for the frame of fig. 4.3; the correct mechanism of collapse must be a special case of one or other of the mechanisms in fig. 4.4. Neither of the two mechanisms of fig. 4.4, moreover, corresponds to a possible equilibrium state of the collapsing

148

---

frame (except for accidental values of the loads leading to a point on the corner of the interaction diagram, such as P or Q in fig. 2.44). This is because, in a frame having three redundancies, the formation of four plastic hinges will completely specify the bending-moment diagram as well as furnishing the collapse equation; the value of the bending moment at the fifth critical section cannot be specified to be fully plastic. This argument may be compared with the writing of four equations (e.g. equations (2.45)) to express collapse of a frame with three redundancies.

Thus the correct solution must involve the elimination of one (at least) of the hinges in fig. 4.4. Examination of the recorded hinge rotations shows that there are three possibilities: $\phi = 0$ (pure sidesway fig. 4.2(a)); $\psi = 0$ (beam mechanism, fig. 4.2(c)); or $\phi = \psi$ (combined mechanism, fig. 4.2(b)). Put another way, if two mechanisms are being combined then they must be 'locked together' if an equilibrium state is to be preserved, and this is done by arranging that a hinge is cancelled.

The 'locking' can be observed numerically as a result of writing the full collapse equation for the general mechanism of fig. 4.4(a). Equating the work done by the loads to the work dissipated in the plastic hinges, as in equation (3.42),

$$\lambda(20\psi + 30\phi) = 15(4\phi + 2\psi),$$

or
$$\lambda = 3\left[\frac{2\phi + \psi}{3\phi + 2\psi}\right]. \tag{4.9}$$

It should be noted that the absolute values of the hinge rotations have been summed to give the total work dissipated in the plastic hinges; equation (4.9) holds, therefore, only for $\phi > \psi$. If $\psi > \phi$, then the mechanism will be as shown in fig. 4.4(b), for which

$$\lambda(20\psi + 30\phi) = 15(2\phi + 4\psi),$$

or
$$\lambda = 3\left[\frac{\phi + 2\psi}{3\phi + 2\psi}\right]. \tag{4.10}$$

As noted in statements (3.50), the value of λ resulting from either equations (4.9) or (4.10) for any assumed values of ϕ and ψ will be an upper bound (unsafe estimate) on the value λ_c at collapse. Thus in equation (4.9), if ϕ is regarded as a constant and ψ allowed to vary, then the least value of λ must be sought. However, since λ is a ratio of linear functions of ψ, it will have no true minimum; to reduce the value of λ, the value of ψ must be made as large as possible. Since $\phi > \psi$ in equation (4.9), the limiting value is $\psi = \phi$, for which the hinge at the left-hand knee disappears, and $\lambda = 1\cdot8$.

Similarly, consideration of equation (4.10) indicates that the value of ψ should be reduced as much as possible, and the limit again occurs for $\psi = \phi$, $\lambda = 1\cdot8$.

It is easy to extend these arguments to show that they are general; mechanisms must be combined so that the resulting mechanism has only a single degree of freedom. The mechanisms may of course be 'incomplete' at any stage, in the sense that the beam mechanism of fig. 4.2(c) leaves the frame once redundant.

The example of fig. 3.16 will be examined by the technique of combining mechanisms; the frame is redrawn in fig. 4.5. Critical sections at which plastic hinges might form must occur at loading points or at joints; there are thus five critical sections for this frame, marked

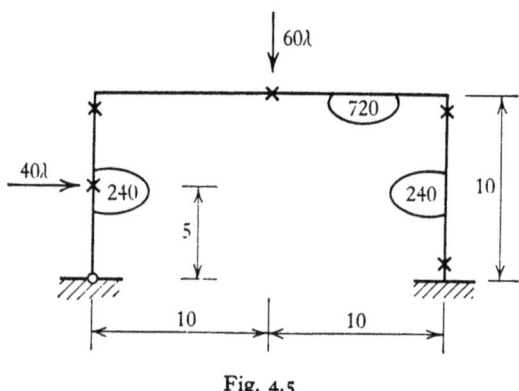

Fig. 4.5

with crosses in fig. 4.5. The frame has two redundancies, and there must therefore be three independent mechanisms of collapse. These may be taken as those sketched in figs. 3.17, 3.18 and 3.19, redrawn in fig. 4.6, and the corresponding collapse equations are

$$
\begin{aligned}
(a) \quad & 200\lambda = 720; \quad \lambda = 3\cdot6, \\
(b) \quad & 200\lambda = 720; \quad \lambda = 3\cdot6, \\
(c) \quad & 600\lambda = 1920; \quad \lambda = 3\cdot2.
\end{aligned}
\tag{4.11}
$$

Examination of the three mechanisms of fig. 4.6 shows that there are only two critical sections (at the top of each column) at which hinges

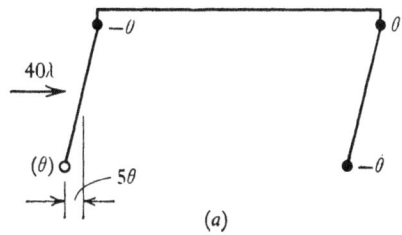

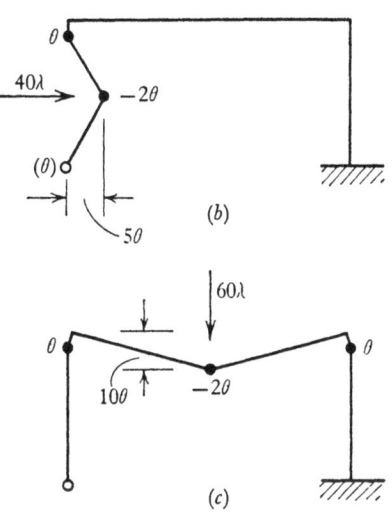

Fig. 4.6

occur in more than one of the mechanisms. For example, the hinge at the right-hand column foot occurs only in fig. 4.6(*a*), and hence cannot be cancelled by combination with any other mechanism. Only the two hinges at the top of the columns can be eliminated in this way. Further, the sway mechanism (from left to right) and the beam mechanisms (collapsing in the direction of the loads) seem entirely reasonable; while the hinge at the top of the right-hand column could be eliminated by subtracting mechanisms (*a*) and (*c*), this would lead to unreasonable mechanisms of the type sketched in fig. 4.7.

It seems likely, therefore (and this assumption will be confirmed later), that the only hinge that can possibly be cancelled by any combination of mechanisms is that at the top of the left-hand column. The negative rotation $(-\theta)$ at this location of fig. 4.6(*a*) can be cancelled *either* with the positive rotation at the same location of fig. 4.6(*c*) *or* with that of fig. 4.6(*b*). The two possibilities are mechanisms (*d*) = (*a*)+(*c*) and (*e*) = (*a*)+(*b*); these mechanisms are those of figs. 3.22 and 3.24, redrawn in fig. 4.8.

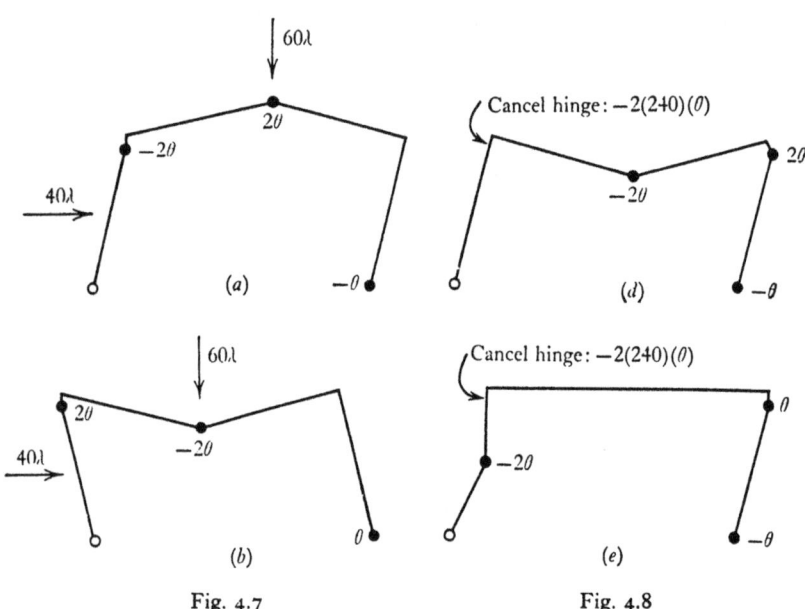

Fig. 4.7   Fig. 4.8

For the first case:

$$
\begin{array}{llll}
(a) & 200\lambda = & 720 \\
(c) & 600\lambda = & 1920 \\
\hline
& 800\lambda & 2640
\end{array}
$$

Cancel hinge: 2(240)

$$(d) \quad 800\lambda = 2160; \quad \lambda = 2\cdot7, \qquad\qquad (4.12)$$

while, for the second case,

$$
\begin{array}{llll}
(a) & 200\lambda = & 720 \\
(b) & 200\lambda = & 720 \\
\hline
& 400\lambda & 1440
\end{array}
$$

Cancel hinge: 2(240)

$$(e) \quad 400\lambda = 960; \quad \lambda = 2\cdot4. \qquad\qquad (4.13)$$

The final result of equation (4.13) is, of course, the correct answer, but it should in general be checked by making a statical analysis (fig. 3.25) of the whole frame to ensure that the yield condition is nowhere violated. For more complex frames it is not so easy to be sure that all possible collapse mechanisms have been investigated, and the static check is an essential part of the analysis.

In this example, the statical analysis of fig. 3.25 confirms that the *mechanism* of fig. 4.8(*e*) gives rise to an *equilibrium* bending-moment distribution which satisfies the *yield* condition. By the uniqueness theorem, the correct value of the load factor has been found, *and no further analysis is necessary*. In particular, there is absolutely no need to investigate mechanisms such as those sketched in fig. 4.7, which were rejected intuitively as being unreasonable modes of collapse for the particular loading system. The correct solution is the correct solution; it is impossible for collapse to occur at a lower load factor by some mode not investigated by the designer.

4.2 Joint rotations

The critical sections at the knees of the frame in fig. 4.5 are marked correctly as lying at the tops of the columns rather than at the ends of the beams; a plastic hinge at a joint between two members will form in

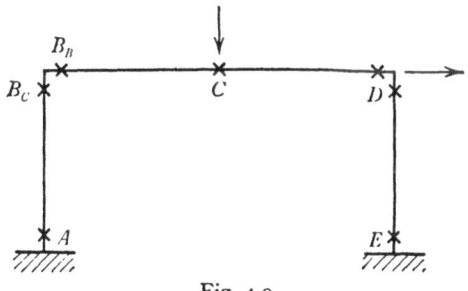

Fig. 4.9

the weaker member. For uniform frames (e.g. fig. 4.2) the hinges at the knees have been shown as lying exactly at the junction of the members, although in practice the hinge must form either in the beam or in the column. To reflect this practical consideration, *two* critical sections should be taken at each knee, fig. 4.9, so that, under the loading shown, the total number of critical sections for the frame is seven. Since the number of redundancies remains at three, then four independent mechanisms are required in order to describe completely the behaviour of the frame.

Four convenient mechanisms are shown in fig. 4.10. Those in fig. 4.10(*a*) and (*b*) are the familiar sway and beam mechanisms; the other two are examples of an important class of elementary mechanism involving *joint rotations*. The meaning of the mechanism in fig. 4.10(*c*),

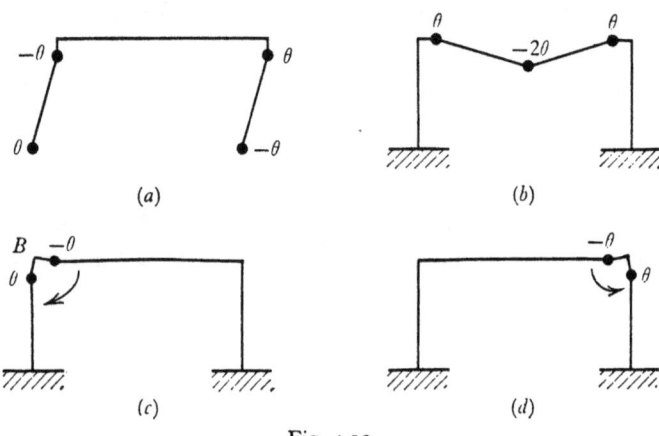

Fig. 4.10

for example, will become clear if the corresponding rotations are used in the equation of virtual work. Denoting by B_C and B_B the cross-sections at joint B lying in the column and beam (fig. 4.9), then the mechanism of fig. 4.10(*c*) leads to the equilibrium equation

$$(M_{B_C})(\theta) + (M_{B_B})(-\theta) = 0,$$

or
$$M_{B_C} = M_{B_B}. \tag{4.14}$$

Equation (4.14) states no more than that the bending moment at knee B of the frame is continuous round the corner.

Such joint rotations are of importance when 'locking together' elementary mechanisms, if the method of combining mechanisms is used. They are essential for the analysis of multi-storey multi-bay frames, but their operation can be seen for the single-bay frame of fig. 4.9, and a discussion may perhaps also help to illuminate the question of cancellation of hinges.

It has been noted that there are only three possible (reasonable) collapse mechanisms for the frame of fig. 4.9, of which two are shown as the independent mechanisms of fig. 4.10(*a*) and (*b*). The third (combined) mechanism results from the addition of these two independent mechanisms, fig. 4.11(*a*); the total hinge rotation remains at 8θ, whereas it will be remembered, equation (4.5), that the total work done in the hinges of the combined mechanism is $(M_p)(6\theta)$. If the left-hand

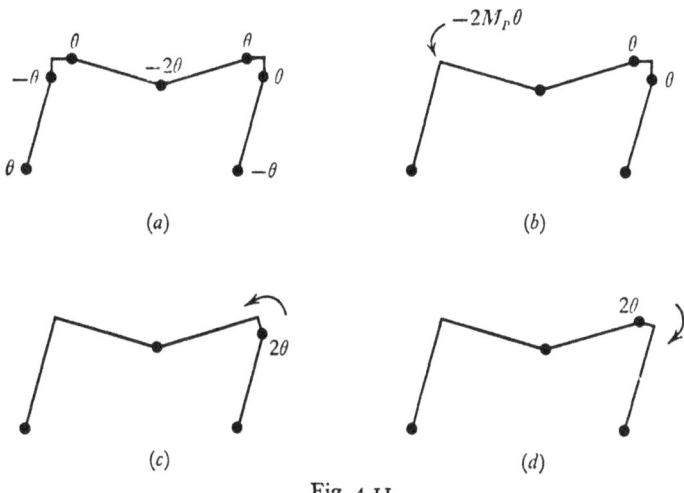

Fig. 4.11

knee of the frame is rotated through θ, then both hinges at the knee will disappear, fig. 4.11(*b*), and the total hinge rotation will be reduced to 6θ. Effectively, the joint rotation of fig. 4.10(*c*) has been used to lock together the two independent mechanisms of collapse, fig. 4.10(*a*) and (*b*), to produce the combined mechanism.

By contrast, it will be seen that rotation of the right-hand knee will not result in any reduction in total hinge rotation at that location. If the beam hinge is closed, for example, as shown in fig. 4.11(*c*), the total hinge rotation at that the knee remains at 2θ, and hence the contribution to the work equation remains at $2M_p\theta$ if the frame has uniform section. Similarly, fig. 4.11(*d*) shows the results of rotating the knee joint through an angle θ in the opposite direction. If the column section differs from that of the beam there will be some advantage, of course, in rotating the joint to locate the plastic hinge in the weaker member.

The use of two critical sections, as at B and D in fig. 4.9, for the case of a joint between two members is perhaps needlessly complicated, but the idea of joint rotations is very useful when the joints are composed of three or more members.

4.3   Multi-storey multi-bay frames

The two-storey single-bay frame of fig. 4.12 has a uniform section, and carries the loads shown. Only one critical section is marked at each end of the upper beam. However, three sections are marked at each end of the lower beam, since a plastic hinge could form either in the beam itself or in the upper or lower column. The bending moments at each

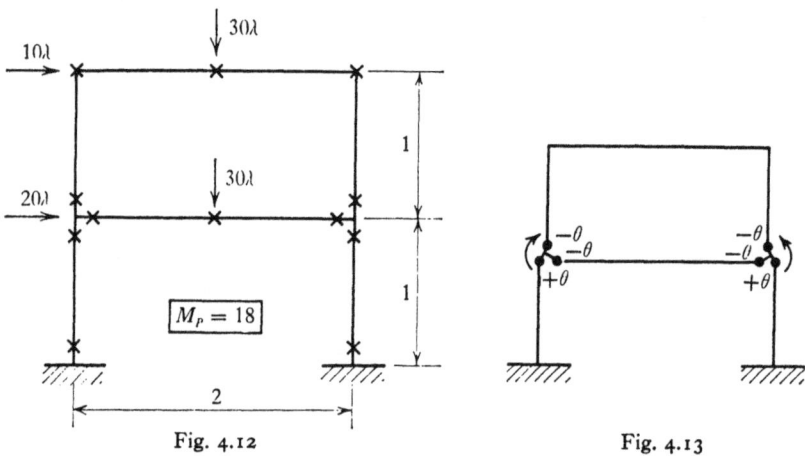

Fig. 4.12   Fig. 4.13

group of three critical sections must sum to zero; the two joint rotations of fig. 4.13 will furnish the corresponding equilibrium equations, similar to equation (4.14).

The signs of the hinge rotations in fig. 4.13 are not really of importance in this simple example; they accord with the previous convention, and with the fuller convention established below (fig. 4.22) for larger frames. So long as a rough sketch is made, it will be clear which way a joint should be rotated so that hinges may be closed up and corresponding deductions made from the terms in the work equation.

The frame of fig. 4.12 has 6 redundancies, and 12 critical sections are marked, so that the number of independent mechanisms is given by the following calculation:

> 12 critical sections (fig. 4.12)
>
> 6 redundancies
> ———————————————
> 6 independent mechanisms
>
> 2 joint rotations (fig. 4.13)
> ———————————————
> 4 'true' mechanisms

It may be noted that had two critical sections been taken at each end of the upper beam, then there would have been 14 critical sections. However, two extra joint rotations would have been introduced, and four true mechanisms would still have resulted.

The four mechanisms, from which may be built up all possible mechanisms of collapse (using, if necessary, the two joint rotations), may be taken conveniently as the simple mechanisms of fig. 4.14. Each beam

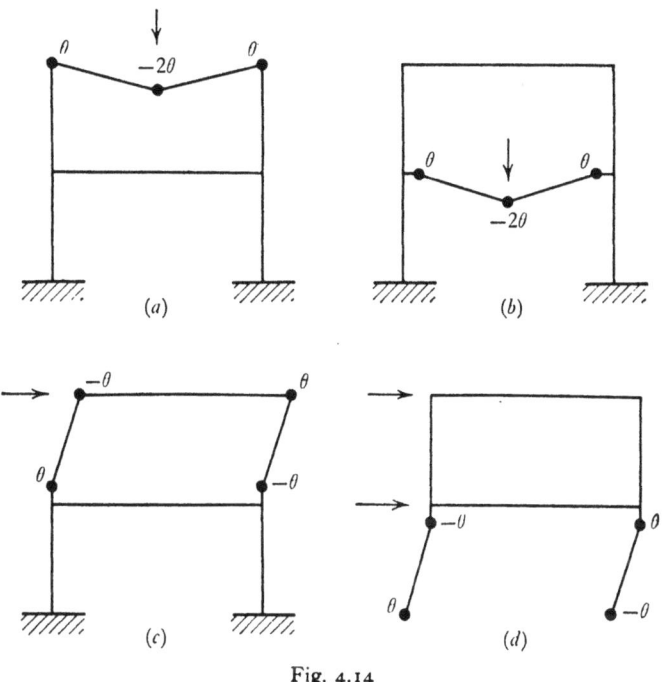

Fig. 4.14

is involved in a local collapse mechanism, and each storey sways in turn. The reader may wish to check that the correct number of 'true' mechanisms can be found in this way for a rectangular building frame having m bays and n storeys; collapse of each beam will give a total of mn mechanisms, and sway of each storey will give n mechanisms, that is, $n(m+1)$ in all. The intermediate totals are given on p. 165.

There are, of course, other possibilities in the choice of four independent mechanisms. For example, had the mechanisms (a), (b) and (c) of fig. 4.14 been chosen, a fourth mechanism, certainly independent of these three, is that illustrated in the second diagram of fig. 4.15. However, the work is usually made easier if the simplest possible mechanisms are used in the first instance.

Using the values of the loads marked in fig. 4.12, the collapse equations corresponding to the independent mechanisms of fig. 4.14 are

$$
\begin{aligned}
(a) \quad & 30\lambda = 72; \quad \lambda = 2\cdot4, \\
(b) \quad & 30\lambda = 72; \quad \lambda = 2\cdot4, \\
(c) \quad & 10\lambda = 72; \quad \lambda = 7\cdot2, \\
(d) \quad & 30\lambda = 72; \quad \lambda = 2\cdot4.
\end{aligned}
\tag{4.15}
$$

These are the four basic equations which, together with the two joint rotations (fig. 4.13) must furnish the correct final solution. The order of combination of the mechanisms is to some extent arbitrary, although it is clearly sensible to start with those elementary mechanisms which give the lowest values of λ. This particular example will be worked twice, with the mechanisms combined in different orders.

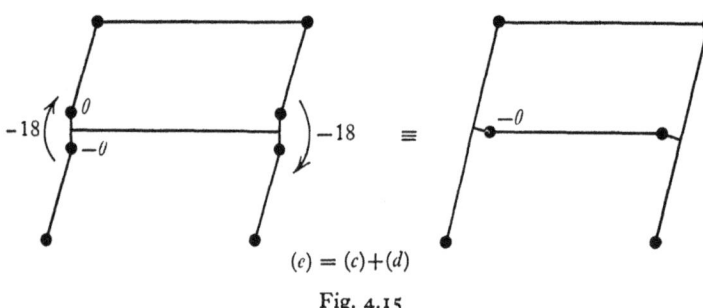

$(e) = (c) + (d)$

Fig. 4.15

Suppose first that the two sidesway mechanisms, (c) and (d) of fig. 4.14, are superimposed, and locked together by means of joint rotations to give a new mechanism (e), fig. 4.15. In this figure it will be seen that the joint rotation at the left-hand end of the lower beam 'closes up' the two hinges in the columns, so that, on this account, $2 \times 'M_p\theta'$ can be subtracted from the right-hand side of the mechanism equations. However, a hinge discontinuity appears in the beam, so that a single '$M_p\theta$' term reappears in the equations. The net 'gain' at the joint is therefore $M_p\theta$, which is numerically equal to eighteen in this example. Put another way, the total hinge discontinuity of 2θ in the left-hand figure of fig. 4.15 is replaced by a discontinuity of θ after the joint rotation.

The same numerical considerations apply to the joint at the right-hand end of the lower beam, and the calculations may be set out:

$$(c) \quad 10\lambda = 72; \quad \lambda = 7\cdot2,$$
$$(d) \quad 30\lambda = 72; \quad \lambda = 2\cdot4,$$
$$\overline{ 40\lambda \quad 144}$$

Rotate joints: $\qquad\qquad\qquad 36$

$$(e) \quad 40\lambda = 108; \quad \lambda = 2\cdot7. \qquad\qquad (4.16)$$

This solution, with $\lambda = 2\cdot7$, is worse than that given by sway of the lower storey alone, but there is now the possibility of adding in the two beam mechanisms, with a corresponding cancellation of hinges. The

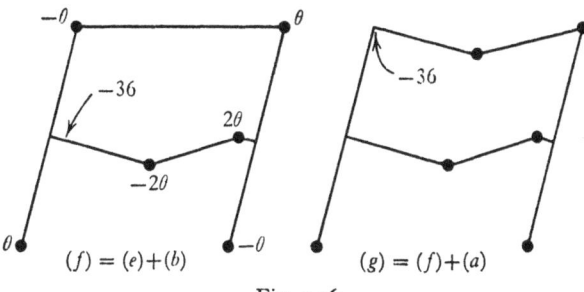

Fig. 4.16

two stages are shown in fig. 4.16. In each case, the cancelled hinge at the left-hand end of the beam gives rise to a deduction of $2M_p\theta$, numerically equal to 36. Starting from equation (4.16) for mechanism (e), the calculations may be set out:

$$(e) \quad 40\lambda = 108; \quad \lambda = 2\cdot7, \qquad\qquad (4.16\ bis)$$
$$(b) \quad 30\lambda = 72$$
$$\overline{ 70\lambda \quad 180}$$

Cancel hinge: $\qquad\qquad\qquad 36$

$$(f) \quad 70\lambda = 144; \quad \lambda = 2\cdot06, \qquad\qquad (4.17)$$
$$(a) \quad 30\lambda = 72$$
$$\overline{ 100\lambda \quad 216}$$

Cancel hinge: $\qquad\qquad\qquad 36$

$$(g) \quad 100\lambda = 180; \quad \lambda = 1\cdot80. \qquad\qquad (4.18)$$

159

4. THE COMBINATION OF MECHANISMS

It would seem that mechanism (g) is correct, since it furnishes the lowest value of the load factor; before making the statical check, however, the same solution will be found by combining the mechanisms in a different order. Fig. 4.17 gives the sequence of operations, and the relevant deductions are shown in each figure; the calculations are:

$$(b) \quad 30\lambda = 72$$
$$(d) \quad 30\lambda = 72$$

$$\overline{\qquad 60\lambda \qquad 144}$$

Rotate joint: $\qquad\qquad\qquad 18$

$$(h) \quad \overline{60\lambda = 126;} \quad \lambda = 2\cdot10, \qquad\qquad (4.19)$$

$$(c) \quad 10\lambda = 72$$

$$\overline{\qquad 70\lambda \qquad 198}$$

Cancel hinge: $\qquad\qquad\qquad 36$

$$\overline{\qquad 70\lambda \qquad 162}$$

Rotate joint: $\qquad\qquad\qquad 18$

$$(f) \quad \overline{70\lambda = 144;} \quad \lambda = 2\cdot06, \qquad\qquad (4.17\ bis)$$

$$(a) \quad 30\lambda = 72$$

$$\overline{\qquad 100\lambda \qquad 216}$$

Cancel hinge: $\qquad\qquad\qquad 36$

$$(g) \quad \overline{100\lambda = 180;} \quad \lambda = 1\cdot80. \qquad\qquad (4.18\ bis)$$

The state of the frame at $\lambda = 1\cdot80$, mechanism (g), is shown in fig. 4.18; the full plastic moments are shown acting at the hinge positions. To demonstrate that the mechanism is correct, the bending moments must be determined at each of the critical sections, and shown to be not greater than $M_p = 18$. An immediate difficulty must be expected, since six hinges only are formed at collapse and the frame has six redundancies; the frame therefore collapses while still being once redundant.

The statical analysis may be made by using the equation of virtual work, or, for this simple problem, equilibrium can be satisfied directly by an examination of individual members of the frame. In fig. 4.19 the full plastic moments have been entered at the hinges, and the bending moments can be computed immediately for the two beams. To complete the analysis of the columns, however, an unknown bending moment M

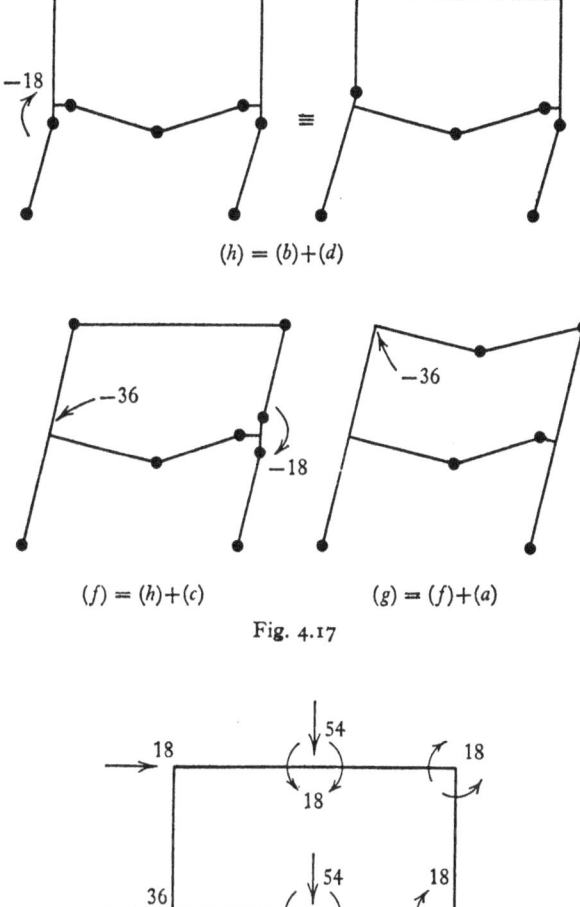

$(h) = (b)+(d)$

$(f) = (h)+(c)$ $(g) = (f)+(a)$

Fig. 4.17

Fig. 4.18

must be introduced. A shear balance across each storey then enables the
column moments to be determined in terms of M.

As expected, the statical analysis has not furnished the value of M;
the frame remains once redundant. The collapse mechanism will be
confirmed as correct, however, if any one value of M can be found such
that the yield condition is satisfied, since this would then complete the

three requirements of the uniqueness theorem. The bending-moment distribution of fig. 4.19 is an equilibrium distribution; it satisfies every test of statics. Second, it corresponds to a mechanism of collapse. Finally, if M is fixed at any value between 0 and 18, the yield condition will be satisfied, so that the load factor $\lambda_c = 1 \cdot 8$ given by mechanism (g), equation (4.18), has been confirmed.

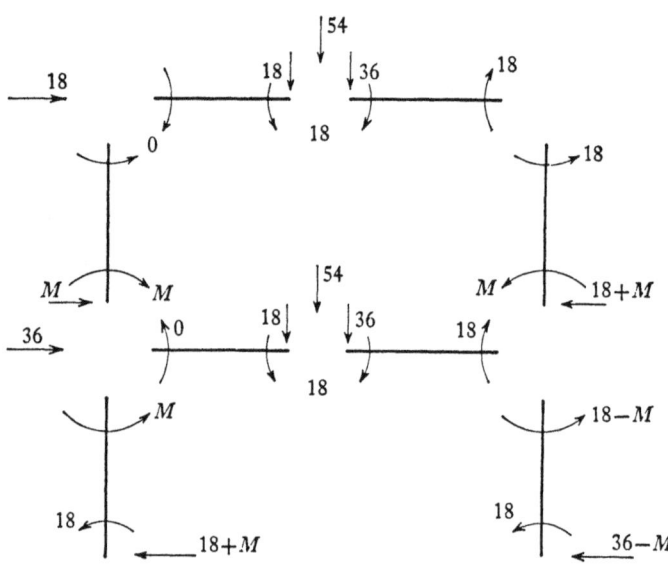

Fig. 4.19

The determination of the values of redundant quantities in a collapsing frame assumes greater importance when the frame is more complex. Methods exist for the determination of suitable values to satisfy the yield condition, but these methods will not be discussed in this volume. For the general design problem, it is essential to obtain some *reasonable* distribution of bending moments in the columns in order to check their, stability; in fig. 4.19, it would not be sufficient for this purpose merely to show that any value of M lying between 0 and 18 satisfies the yield condition. However, the determination of column bending moments is in turn part of a wider problem in which attention must be paid to the disposition of dead and live load on the floors of the building.

As another example of analysis by combination of mechanisms, the frame of fig. 4.20 will be examined. In this frame the beam loads are shown as uniformly distributed of the magnitudes marked above each beam; the full plastic moments are shown against each member. The uniformly distributed loads are realistic from a practical point of view, but an infinite number of critical sections (at which hinges might form) is possible in each beam. The question of distributed loads is dealt with

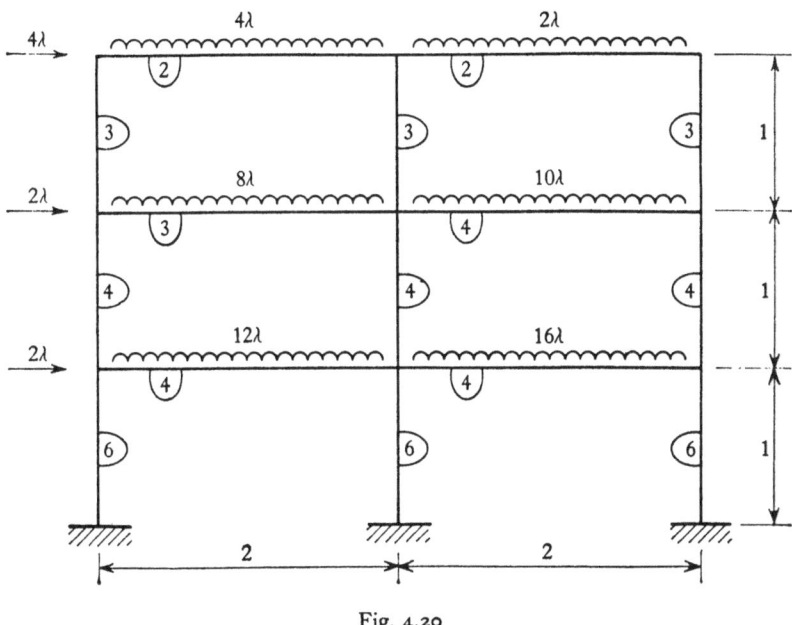

Fig. 4.20

below (section 4.4); for the time being, it may be noted that, in the preliminary analysis of such a frame, a close estimate of the collapse load factor will be obtained if only a single critical section is taken at the centre of length of each beam.

Now the maximum free bending moment for a simply supported beam under uniformly distributed loading is '$Wl/8$'; the corresponding value for a central point load is '$Wl/4$'. Thus each uniformly distributed load may be replaced by a central point load of half the value to give the same free bending moment at the central critical section. The frame of fig. 4.21 will therefore be analysed, the full plastic moments of the members being as shown in fig. 4.20.

In the pictorial addition of mechanisms (as in fig. 4.17 for example) there is no real need for a sign convention for the bending moments.

4. THE COMBINATION OF MECHANISMS

However, it is convenient (in using virtual work, for example, or in setting up a program for automatic computation) to have a consistent convention; almost any convention will do, so long as it is adhered to

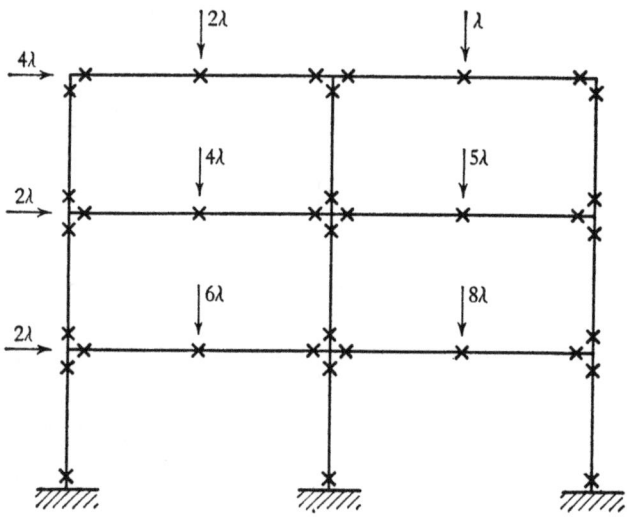

Fig. 4.21

rigidly. A system which has some advantages of simplicity is shown in fig. 4.22; bending moments producing compression on faces of members adjacent to the broken lines will be denoted positive. This convention has already been used for single- and two-storey frames, and is readily extensible for extra storeys and bays.

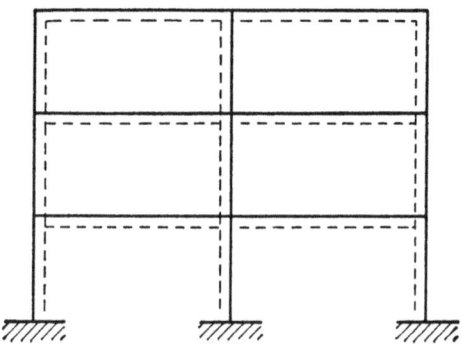

Fig. 4.22

The frame in fig. 4.21 has thirty-six critical sections:

36 critical sections	$[5mn + 2n]$
18 redundancies	$[3mn]$
18 independent mechanisms	$[2n(m+1)]$
9 joints	$[n(m+1)]$
9 true mechanisms	$[nm$ beams $+n$ sways$]$

The 9 mechanisms (6 beams and 3 sways) are shown in fig. 4.23, and lead to the following collapse equations:

$$
\begin{aligned}
(a) \quad & 2\lambda = 8, \quad \lambda = 4\cdot00, \\
(b) \quad & \lambda = 8, \quad \lambda = 8\cdot00, \\
(c) \quad & 4\lambda = 12, \quad \lambda = 3\cdot00, \\
(d) \quad & 5\lambda = 12, \quad \lambda = 2\cdot40, \\
(e) \quad & 6\lambda = 16, \quad \lambda = 2\cdot67, \\
(f) \quad & 8\lambda = 16, \quad \lambda = 2\cdot00, \\
(g) \quad & 4\lambda = 18, \quad \lambda = 4\cdot50, \\
(h) \quad & 6\lambda = 24, \quad \lambda = 4\cdot00, \\
(j) \quad & 8\lambda = 36, \quad \lambda = 4\cdot50.
\end{aligned}
\qquad (4.20)
$$

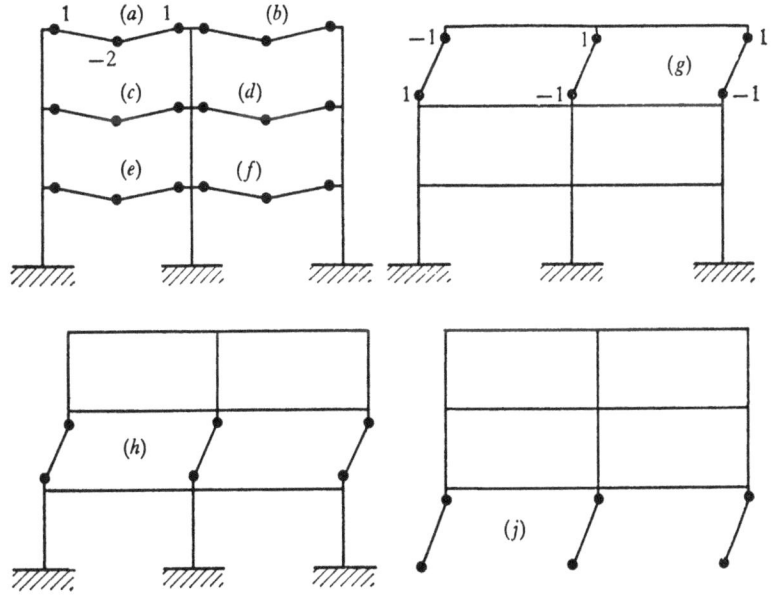

Fig. 4.23

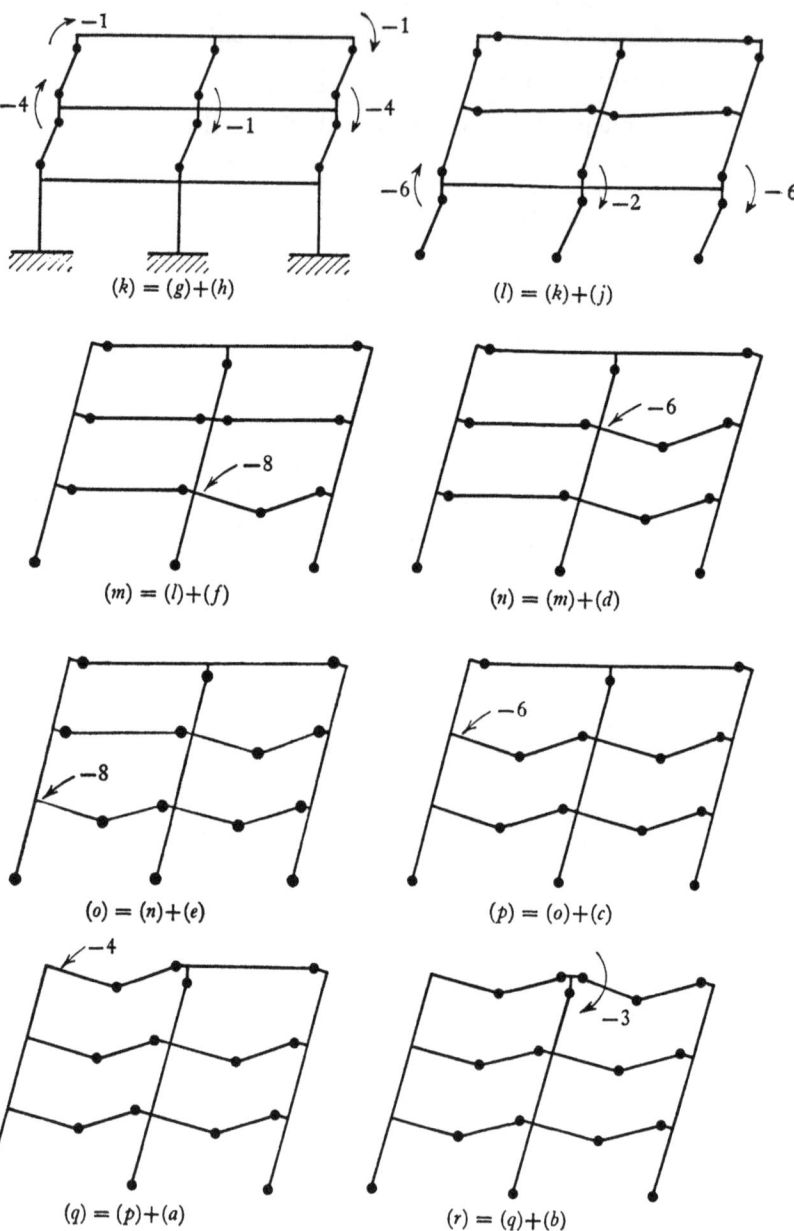

Fig. 4.24

The beam mechanism (f) gives the lowest value of the load factor, and it is possible (although not actually so in this case) that a single beam collapse might be the correct collapse mode for the frame as a whole. In any case, it seems likely that mechanism (f) will be contained in any derived collapse mode, and it must be investigated whether this beam can be combined with any of the sway mechanisms (and other beams) to give a lower value of the load factor.

The sequence of operations sketched in fig. 4.24 will be followed, in which the three sway mechanisms are first used, followed by the beam mechanisms in ascending order of their individual load factors. The calculations may be laid out as before:

$$(g) \quad 4\lambda = 18$$
$$(h) \quad 6\lambda = 24$$
$$\overline{\quad\quad 10\lambda \quad 42}$$

Rotate joints: $\quad\quad\quad\quad\quad 11$

$$(k) \quad 10\lambda = 31; \quad \lambda = 3\cdot10, \quad\quad\quad (4.21)$$
$$(j) \quad 8\lambda = 36$$
$$\overline{\quad\quad 18\lambda \quad 67}$$

Rotate joints: $\quad\quad\quad\quad\quad 14$

$$(l) \quad 18\lambda = 53; \quad \lambda = 2\cdot94, \quad\quad\quad (4.22)$$
$$(f) \quad 8\lambda = 16$$
$$\overline{\quad\quad 26\lambda \quad 69}$$

Cancel hinge: $\quad\quad\quad\quad\quad 8$

$$(m) \quad 26\lambda = 61; \quad \lambda = 2\cdot35, \quad\quad\quad (4.23)$$
$$(d) \quad 5\lambda = 12$$
$$\overline{\quad\quad 31\lambda \quad 73}$$

Cancel hinge: $\quad\quad\quad\quad\quad 6$

$$(n) \quad 31\lambda = 67; \quad \lambda = 2\cdot16, \quad\quad\quad (4.24)$$
$$(e) \quad 6\lambda = 16$$
$$\overline{\quad\quad 37\lambda \quad 83}$$

Cancel hinge: $\quad\quad\quad\quad\quad 8$

$$(o) \quad 37\lambda = 75; \quad \lambda = 2\cdot03, \quad\quad\quad (4.25)$$

$$(c), \quad 4\lambda = 12$$

$$\overline{\quad 41\lambda \qquad 87}$$

Cancel hinge: $\qquad\qquad\qquad 6$

$$(p) \quad 41\lambda = 81; \quad \lambda = 1\cdot976, \qquad\qquad (4.26)$$

$$(a) \quad 2\lambda = 8$$

$$\overline{\quad 43\lambda \qquad 89}$$

Cancel hinge: $\qquad\qquad\qquad 4$

$$(q) \quad 43\lambda = 85; \quad \lambda = 1\cdot977, \qquad\qquad (4.27)$$

$$(b) \qquad \lambda = 8$$

$$\overline{\quad 44\lambda \qquad 93}$$

Rotate joint: $\qquad\qquad\qquad 3$

$$(r) \quad 44\lambda = 90; \quad \lambda = 2\cdot05. \qquad\qquad (4.28)$$

It will be seen that until mechanism (p), equation (4.26), for which $\lambda = 1\cdot976$, the load factor has not been reduced below the original lowest value, $\lambda = 2$ for mechanism (f). The incorporation of the roof beams into the mechanism increases the value of the load factor, and the conclusion from this analysis is that mechanism (p) is probably correct. The confirmation by a statical analysis is left to the reader; it should be noted that the frame had eighteen redundancies, and only fourteen hinges are formed in mechanism (p), so that five redundancies will be left at collapse.

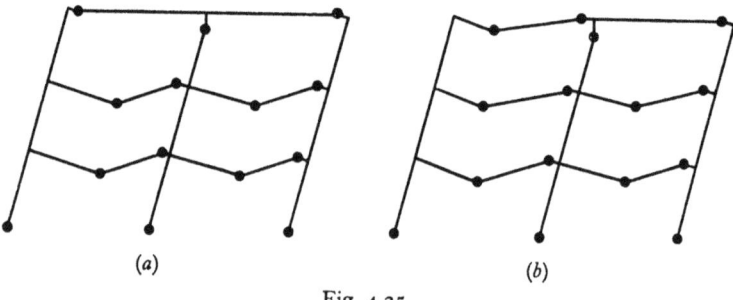

(a)   (b)

Fig. 4.25

The collapse mode is redrawn in fig. 4.25(a). This mode can be modified to allow for the effect of the actual loads being uniformly distributed, and an exact analysis leads to the mode sketched roughly in fig. 4.25(*b*), in which hinges form away from the mid-points of the beams. The exact value of the load factor is 1·94 instead of the value 1·98 found above for mechanism (*p*).

4.4 Distributed loads

There is no case of practical importance for which it would be necessary to make exact calculations for the effect of distributed loads, and, as mentioned above, a single critical section at the centre of each beam is usually all that is necessary. The single-bay frame of uniform section shown in fig. 4.26 will first be solved exactly; an analysis will then be made with critical sections allowed only at the quarter-points and the mid-point of the beam.

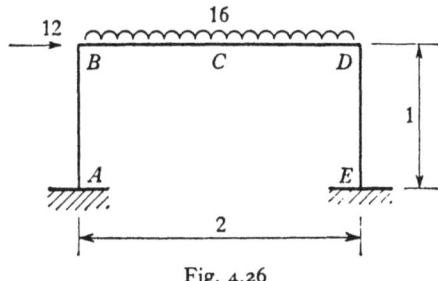

Fig. 4.26

The pure sidesway mode of collapse (fig. 4.29(*a*)) would lead to the equation

$$4M_p = 12,$$

i.e.

$$M_p = 3·00, \qquad (4.29)$$

while the beam mechanism, fig. 4.29(*c*), gives

$$4M_p = 8,$$

or

$$M_p = 2·00. \qquad (4.30)$$

In fact, the correct mechanism is that sketched in fig. 4.27, where the location of the sagging hinge in the beam is as yet unknown; the collapse equation is

$$(12)(\theta) + (16)(\tfrac{1}{2}z\theta) = (M_p)\left[2 + 2\left(\frac{2}{2-z}\right)\right]\theta,$$

or

$$M_p = \frac{2(6 + z - 2z^2)}{4 - z}. \qquad (4.31)$$

4. THE COMBINATION OF MECHANISMS

The condition for M_p to be a maximum is

$$(1 - 4z)(4 - z) + (6 + z - 2z^2) = 0,$$

or

$$z = 4 - \sqrt{11} = 0.683, \tag{4.32}$$

and the corresponding value of M_p is given by

$$M_p = 30 - 8\sqrt{11} = 3.47. \tag{4.33}$$

The reader may wish to confirm this solution by sketching the bending-moment diagram subject to the conditions (4.32) and (4.33).

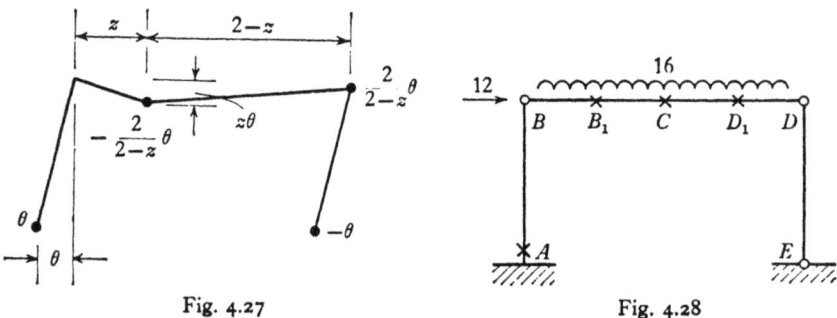

Fig. 4.27   Fig. 4.28

For the approximate solution, 7 critical sections will be taken, including the points B_1 and D_1 at quarter-span marked in fig. 4.28. In this figure, the frame has been made statically determinate by the insertion of three pins, and the corresponding table of free moments may be drawn up:

Table 4.1

	A	B	B_1	C	D_1	D	E
Free moment:	12	0	−3	−4	−3	0	0

Since the frame has three redundancies, there will be four independent mechanisms; the two usual mechanisms are shown in fig. 4.29(a) and (c), while two more mechanisms, involving hinges at B_1 and D_1, are sketched in fig. 4.29(b) and (d). The collapse equations for these four mechanisms will be written, using equation (3.11),

$$\Sigma(M_F)\phi = \Sigma M_p |\phi|, \tag{4.34}$$

where the values of the free moments M_F are taken from table 4.1 above. Thus the four fundamental collapse equations are

$$
\begin{aligned}
(a) \quad & 12 = 4M_p; \quad M_p = 3\cdot00, \\
(b) \quad & 12 = 8M_p; \quad M_p = 1\cdot50, \\
(c) \quad & 8 = 4M_p; \quad M_p = 2\cdot00, \\
(d) \quad & 12 = 8M_p; \quad M_p = 1\cdot50.
\end{aligned}
\qquad (4.35)
$$

These mechanisms must be combined in an attempt to achieve a value of M_p greater than $3\cdot00$ (for mechanism (a) above). The only common hinge in fig. 4.29 that can be cancelled is that at the left-hand knee B, and each of the three beam mechanisms will be tried in turn with the sway mechanism.

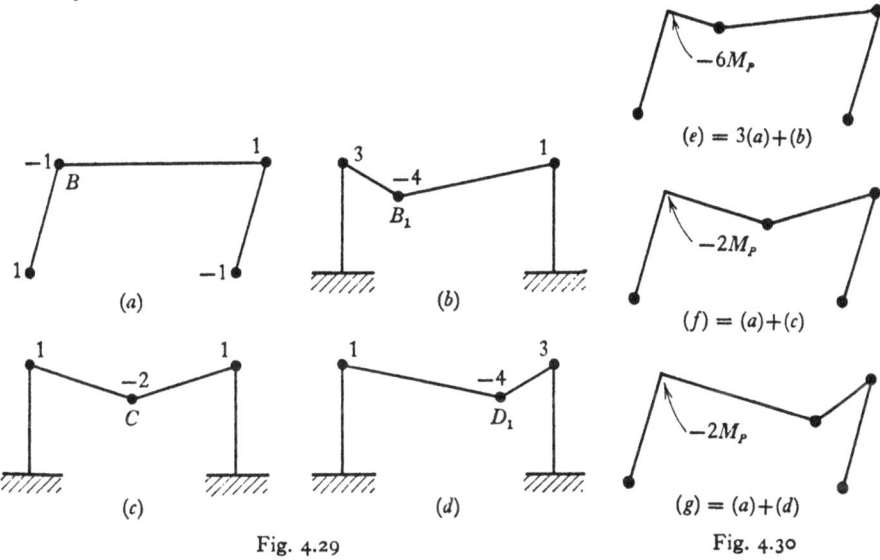

Fig. 4.29   Fig. 4.30

As sketched in fig. 4.29, the hinge rotations at B for mechanisms (a) and (b) are not the same, and hence cannot be cancelled by direct superposition. If, however, the hinge rotations of mechanism (a) were increased threefold, then the hinge could be cancelled. This 'scaling' of a mechanism must be reflected in the writing of the corresponding equation; that is, equation 4.35(a) must be multiplied by a factor 3 throughout before adding it to equation 4.35(b). The combination of these two mechanisms results in mechanism (e) of fig. 4.30; note the reduction of $6M_p$ due to the cancellation of the hinge, since each hinge rotation at B is three units. The calculations, together with those for the other two cases of fig. 4.30, may be laid out as follows:

$$3\,(a) \quad 36 = 12M_p$$
$$(b) \quad 12 = 8M_p$$
$$\overline{\quad 48 \quad 20M_p}$$

Cancel hinge: $\qquad\qquad\qquad 6M_p$

$$(e) \quad \overline{48 = 14M_p}; \quad M_p = 3{\cdot}43. \qquad\qquad (4.36)$$

$$(a) \quad 12 = 4M_p$$
$$(c) \quad\ \ 8 = 4M_p$$
$$\overline{\quad 20 \quad 8M_p}$$

Cancel hinge: $\qquad\qquad\qquad 2M_p$

$$(f) \quad \overline{20 = 6M_p}; \quad M_p = 3{\cdot}33. \qquad\qquad (4.37)$$

$$(a) \quad 12 = 4M_p$$
$$(d) \quad 12 = 8M_p$$
$$\overline{\quad 24 \quad 12M_p}$$

Cancel hinge: $\qquad\qquad\qquad 2M_p$

$$(g) \quad \overline{24 = 10M_p}; \quad M_p = 2{\cdot}40. \qquad\qquad (4.38)$$

Evidently, mechanism (e), equation (4.36), gives the correct approximate solution, and it should be noted that the corresponding value of M_p differs by little more than 1 % from the exact value, equation (4.33). Had only a single critical section been taken, at the centre of the beam, then the approximate solution, equation (4.37), would still have differed by only 4 % from the exact solution.

4.5 The pitched-roof frame

The pitched-roof frame of fig. 2.61, redrawn in fig. 4.31, will be analysed once more, using this time the technique of combining mechanisms. As a first approximate solution, plastic hinges will be permitted only at the five cardinal sections A, B, C, D and E. Since the frame has three redundancies, there will be two basic independent mechanisms of collapse, which may well be taken as (a) and (c) of fig. 2.57, redrawn in fig. 4.32. The hinge rotations of fig. 4.32(a) may be established by referring to the general values shown in fig. 2.58.

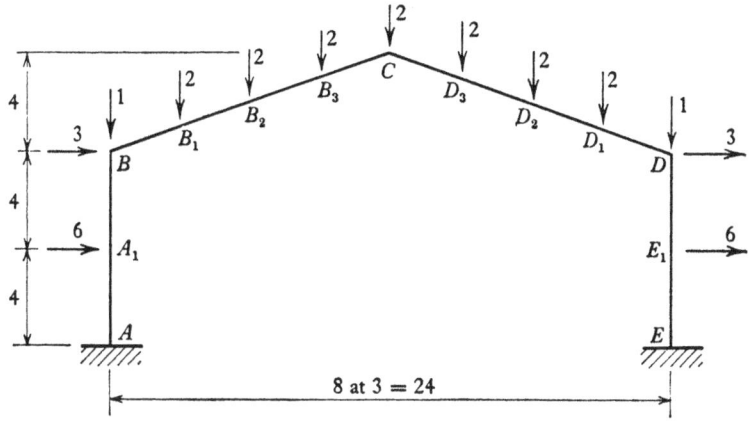

Fig. 4.31

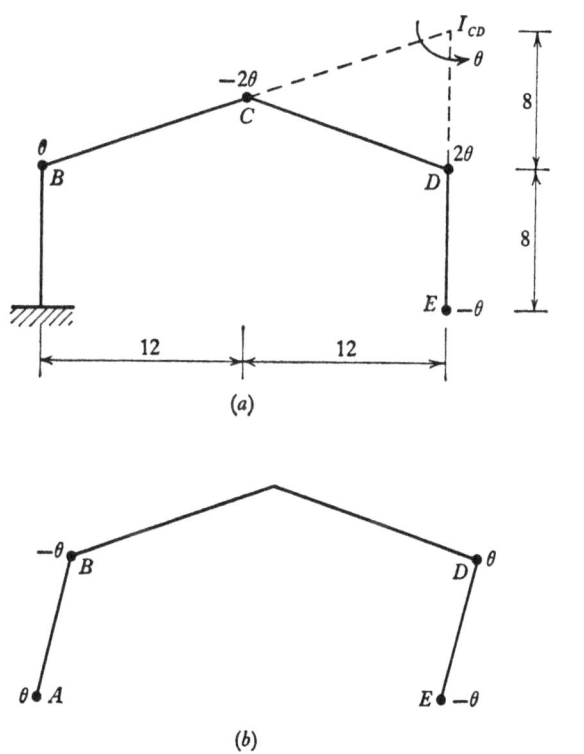

(a)

(b)

Fig. 4.32

173

4. THE COMBINATION OF MECHANISMS

The virtual work equation in the form (4.34) will again be used to establish the basic mechanism equations; the free bending moments of table 2.3 are retabulated in table 4.2:

Table 4.2

	A	A_1	B	B_1	B_2	B_3	C	D_3	D_2	D_1	D	E_1	E
Free:	96	60	48	27	12	3	0	3	12	27	48	36	0

Thus the two basic collapse equations corresponding to the mechanisms of fig. 4.32 are

$$(a) \quad 144 = 6M_p; \quad M_p = 24, \qquad (4.39)$$

$$(b) \quad 96 = 4M_p; \quad M_p = 24. \qquad (4.40)$$

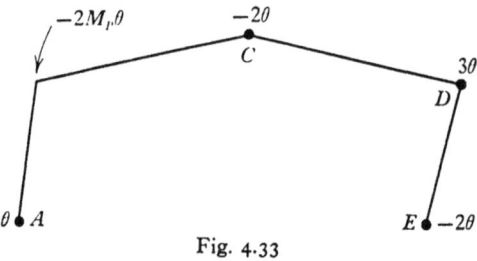

Fig. 4.33

Combination of these two mechanisms is illustrated in fig. 4.33, where the hinge rotations are simply the sums of those in the two mechanisms of fig. 4.32 (cf. fig. 3.11). Using equations (4.39) and (4.40),

$$(a) \quad 144 = 6M_p$$
$$(b) \quad 96 = 4M_p$$
$$\overline{ 240 \quad 10M_p}$$

Cancel hinge B:
$$\overline{ 2M_p}$$

$$(\text{fig. } 4.33) \quad 240 = 8M_p; \quad M_p = 30. \qquad (4.41)$$

So far, then, by the unsafe theorem, the value of M_p must be at least 30; this value is very close to the true value of 30·75, equations (2.72) and table 2.10. The exact value can be found by allowing plastic hinges to form at the actual critical sections at the sheeting rails (A_1 and E_1) and the purlin points (B_1, etc.).

174

Now the number of independent mechanisms required for the analysis of a frame is equal to the number of critical sections less the number of redundancies of the frame. For a given frame, therefore, whose number of redundancies is fixed, the specification of an extra critical section, say B_3 in fig. 4.31, will require the construction of an extra basic independent mechanism (which must, of course, involve a hinge rotation at the new critical section). Thus the local collapse mechanism of fig. 4.34 is a possible extra independent mechanism involving the critical section B_3; similarly, the local collapse mechanism

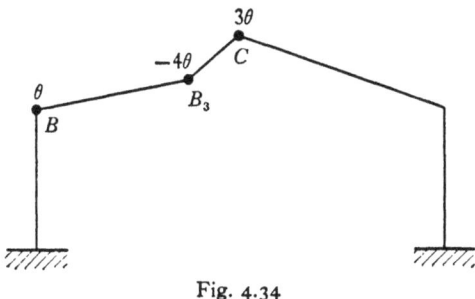

Fig. 4.34

of fig. 4.37 involves the section B_2. A similar additional mechanism must be taken for each extra critical section that it is desired to introduce into the analysis.

The collapse equation for the mechanism of fig. 4.34 may be written, again using equation (4.34) and table 4.2 of free moments,

$$\text{(fig. 4.34)} \quad 36 = 8M_p; \quad M_p = 4 \cdot 5. \tag{4.42}$$

To combine the mechanisms of figs. 4.33 and 4.34 the hinge rotations at C must first be made the same; that is, the rotations of fig. 4.33 (and the numbers in equation (4.41)) should be multiplied by a factor of 3, and those of fig. 4.34 and equation (4.42) by a factor of 2, in order that the hinge at C be cancelled. Thus the calculations may be written in the usual form:

$$3 \text{ (fig. 4.33)} \quad 720 = 24M_p$$
$$2 \text{ (fig. 4.34)} \quad \underline{72 = 16M_p}$$
$$792 \qquad 40M_p$$
Cancel hinge C: $\qquad\qquad\quad 12M_p$
$$\text{(fig. 4.35)} \quad 792 = 28M_p; \quad M_p = 28 \cdot 3. \tag{4.43}$$

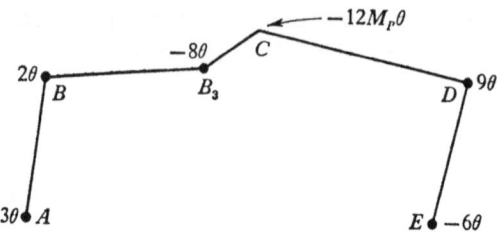

Fig. 4.35

The mechanism of fig. 4.35 has five hinges, and hence cannot correspond to an equilibrium state. However, hinge B can be 'closed up' by
using again the sway mechanism of fig. 4.32:

$$\text{(fig. 4.35)} \qquad 792 = 28M_p \qquad\qquad \text{(4.43 bis)}$$
$$2\text{ (fig. 4.32}(b)) \quad 192 = 8M_p$$
$$\overline{984 \qquad 36M_p}$$

Cancel hinge B: $\qquad\qquad\qquad\qquad\qquad 4M_p$

$$\text{(fig. 4.36)} \qquad 984 = 32M_p; \quad M_p = 30\cdot75. \quad \text{(4.44)}$$

The mechanism of fig. 4.36 is admissible, having only four hinges, and
is, in fact, correct (cf. fig. 3.12); it gives the largest value of M_p.

An alternative possibility for the correct collapse mechanism (which
might occur for higher vertical and smaller horizontal loading than in
the present example) is shown in fig. 4.39, which can be developed from
fig. 4.33 by using the independent mechanism of fig. 4.37:

$$\text{(fig. 4.33)} \qquad 240 = 8M_p \qquad\qquad \text{(4.41 bis)}$$
$$2\text{ (fig. 4.37)} \qquad 48 = 8M_p; \quad M_p = 6.$$
$$\overline{288 \qquad 16M_p}$$

Cancel hinge C: $\qquad\qquad\qquad\qquad\qquad 4M_p$

$$\text{(fig. 4.38)} \qquad 288 = 12M_p; \quad M_p = 24. \qquad \text{(4.45)}$$
$$2\text{ (fig. 4.32}(b)) \quad 192 = 8M_p$$
$$\overline{480 \qquad 20M_p}$$

Cancel hinge B: $\qquad\qquad\qquad\qquad\qquad 4M_p$

$$\text{(fig. 4.39)} \qquad 480 = 16M_p; \quad M_p = 30. \qquad \text{(4.46)}$$

The value $M_p = 30$ is lower than that of equation (4.44), confirming
that the mode of fig. 4.39 is not the correct collapse mode.

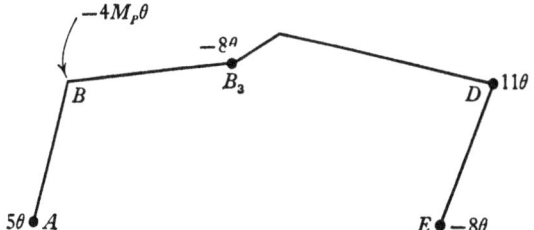

Fig. 4.36

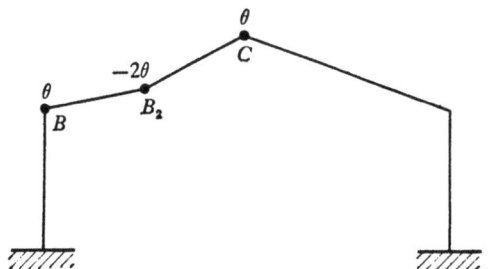

Fig. 4.37

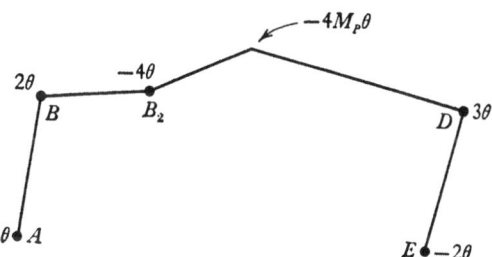

Fig. 4.38

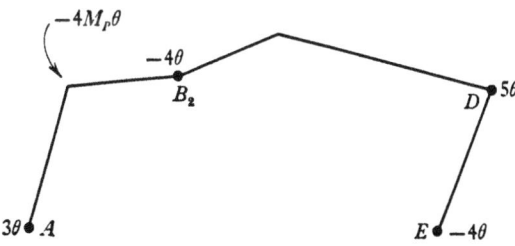

Fig. 4.39

4. THE COMBINATION OF MECHANISMS

Other possible mechanisms of collapse can be examined in a similar way, but the method of combination of mechanisms is in fact cumbersome for this particular example, as it is for any frame having a relatively small number of redundancies and a relatively large number of loading points (i.e. potential critical sections). For such frames, the 'direct' methods of chapter 2 will generally lead to a simpler analysis, particularly if free bending-moment diagrams are sketched (e.g. fig. 2.63 for the pitched-roof frame); physically unlikely mechanisms can be spotted from such diagrams and excluded from the calculations.

EXAMPLES

4.1 The fixed-base frame shown is of uniform cross-section, full plastic moment 100 kNm. Determine the collapse load factor. (*Ans.* 16/9 = 1·78.)

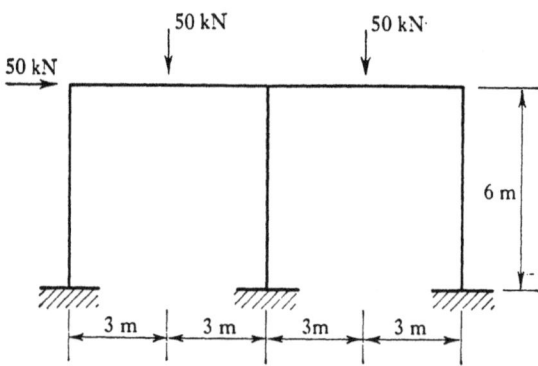

4.2 Find the full plastic moment required for the frame shown. Columns and beams have the same section. (*Ans.* 150 kNm.)

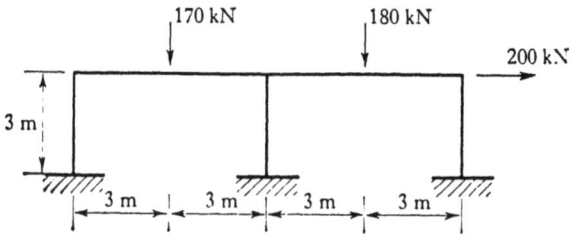

EXAMPLES

4.3 Find the collapse load factor of the frame shown. The full plastic moments of the members (kNm) are marked in the figure. (*Ans.* 1·48.)

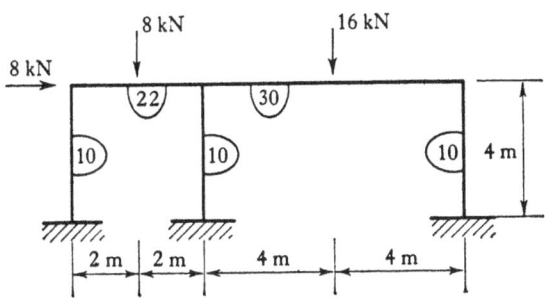

4.4 The frame shown has uniform section, full plastic moment 45 kNm. Determine the collapse load factor. (*Ans.* 2·00.)

4.5 Repeat example 4.4 if the load of 10 kN acting on the top beam is replaced by a load of (*a*) 15 kN, (*b*) 20 kN. (*Ans.* 2·00, 1·875.)

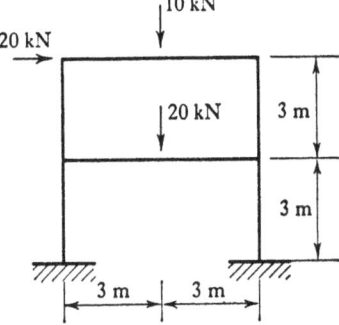

4.6 The frame shown carries working loads of the values indicated; the full plastic moments of the members, in kNm, are also shown in the figure. All connexions are full strength. Making the usual assumptions of simple plastic theory, find the load factor against collapse of the frame. (*Ans.* 19/9 = 2·11.)

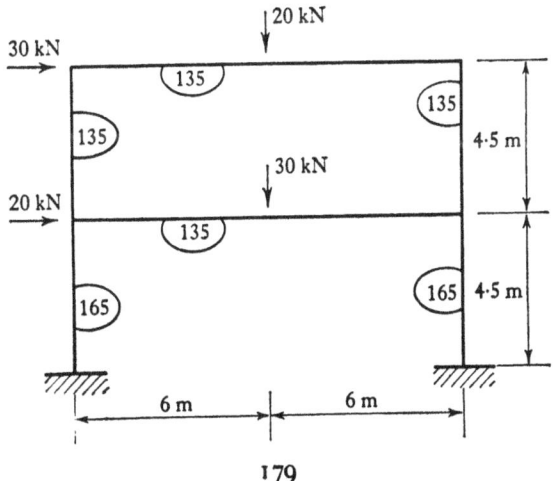

179

4. THE COMBINATION OF MECHANISMS

The solution will not be considered complete unless a satisfactory bending-moment distribution at collapse is determined. (*M.S.T.* II, 1961; adapted.)

(*Ans.* 2·11.)

4.7 The portal frame shown is to be designed to have uniform section of full plastic moment M_p; the feet are fixed and all connexions between members are full strength. The loads W act at mid-span of the beams.

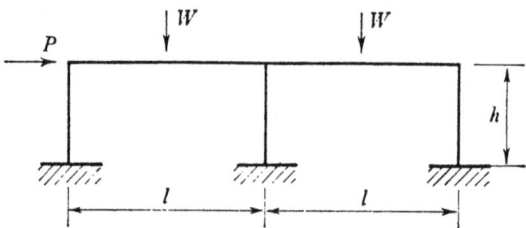

Using axes Ph/M_p, Wl/M_p, construct an interaction diagram from which the mode of collapse and the value of M_p required may be determined for all ratios Ph/Wl. It may be assumed that the values of P and W incorporate a suitable load factor, and that both are positive.

Hence, or otherwise, find the value of M_p required for $h = 3$ m, $l = 6$ m, $P = 50$ kN, and $W = 30$ kN. (*M.S.T.* II, 1962; adapted.)

(*Ans.* 30 kNm.)

4.8 For the plane two-storey portal frame with rigid joints and encastré stanchion feet shown, the ratio of the full plastic moment of resistance of the stanchions to that of the beams is 0·8.

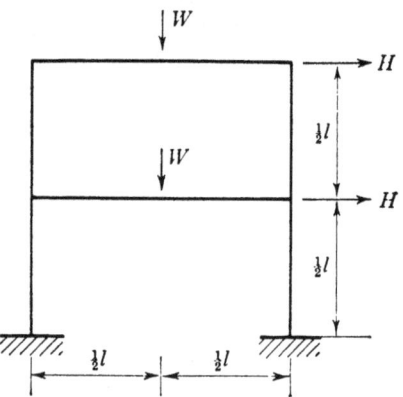

Investigate the plastic collapse behaviour of the frame for various ratios $W:H$ of the vertical and horizontal concentrated loads shown, and present the results in graphical form. The usual simplifying assumptions of the plastic theory may be used. (*M.S.T.* II, 1958.)

4.9 The frame shown has uniform section, full plastic moment 10 units. Determine the collapse load factor. (*Ans.* 2·50.)

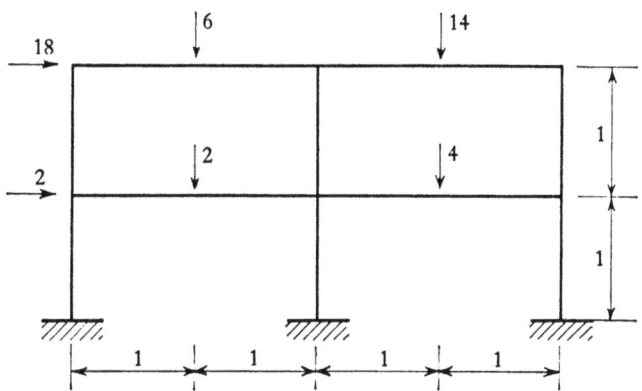

4.10 The two-bay two-storey frame shown has full-strength connexions and carries the working loads shown; the full plastic moments are marked adjacent to the members.

Determine the collapse load factor according to simple plastic theory. (*M.S.T.* II, 1968.) (*Ans.* 15/7).

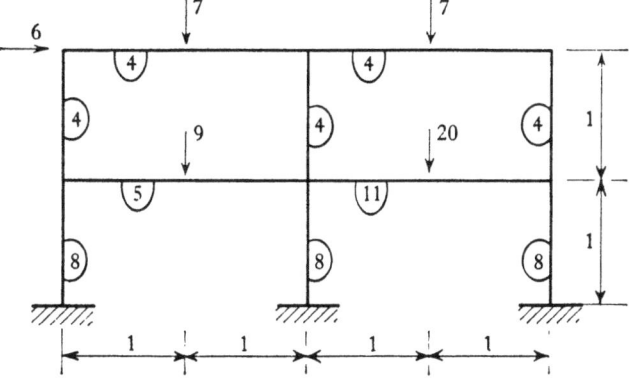

4. THE COMBINATION OF MECHANISMS

4.11 Making the usual assumptions of the simple plastic theory, find a good upper bound on the value of W to cause collapse of the frame shown. The vertical loads act at the midpoints of the beams, and the full plastic moments are marked in the figure against each member. (*M.S.T.* II, 1965.)

(*Ans.* $3 \cdot 60 \ M_p/l$ (exact).)

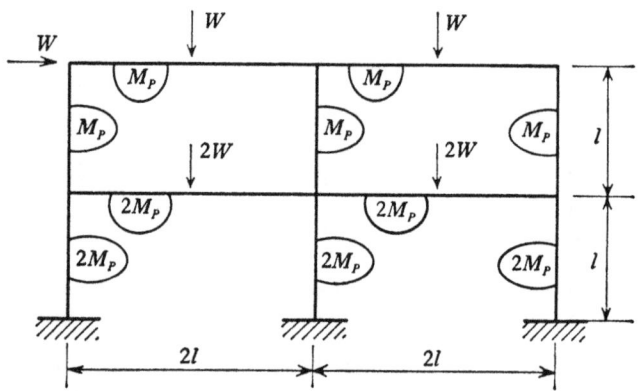

4.12 The frame shown is composed of members with the following uniform full plastic moments in kNm: $AB = 200$, $BC = 150$, $DE = 500$, $EF = 400$, $AD = 50$, $BE = 65$, $CF = 50$, $DG = 130$, $EH = 200$, $FI = 130$. Determine the collapse load factor. (*M.S.T.* II, 1956; adapted.) (*Ans.* $1 \cdot 673$.)

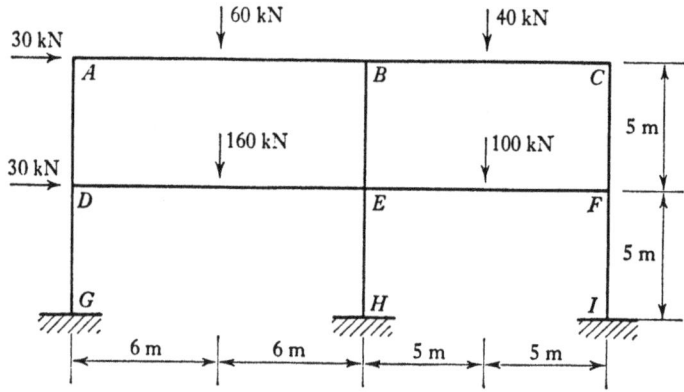

4.13 In the welded frame shown in which the feet are fixed, the beams have length 2 units and each storey has height 1 unit. The members have the full plastic moments indicated. Determine the factor by which the loads shown must be multiplied in order that collapse should just occur. The usual assumptions of simple plastic theory should be made. (*M.S.T.* II, 1967.)

(*Ans.* 2·00.)

4. THE COMBINATION OF MECHANISMS

4.14 In the four-storey frame shown, the loads are in kN. The two upper beams have full plastic moment 531 kNm, and the two lower beams have full plastic moment 583 kNm. Both columns are of uniform section throughout the four storeys, and have full plastic moment 355 kNm. The feet of the frame are fixed, and all connexions are full strength.

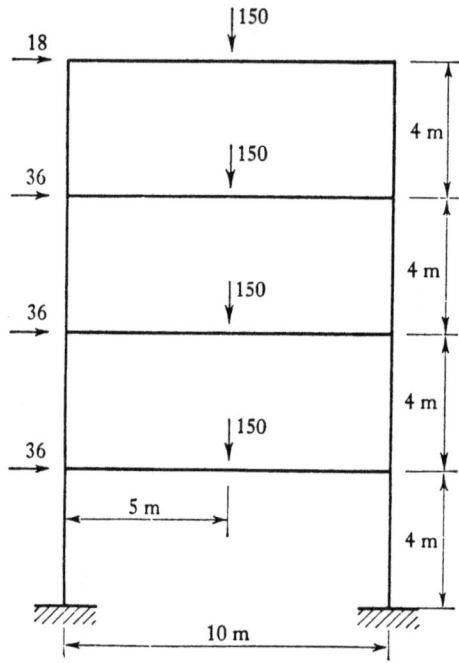

Making the usual assumptions of simple plastic theory, determine the load factor against collapse of the frame. (*M.S.T.* II, 1962; adapted.)

(*Ans.* 2·23.)

4.15 The members of the rigid frame shown have the relative plastic moments shown ringed, and the frame is to collapse under the loading shown. Assuming, as a first approximation, that plastic hinges occur either at the joints or at the mid-span of the members, and neglecting the effects of axial load and instability, determine the required full plastic moments. (Leeds University 1963; adapted.)   (*Ans.* $M_p = 53 \cdot 2$ kNm.)

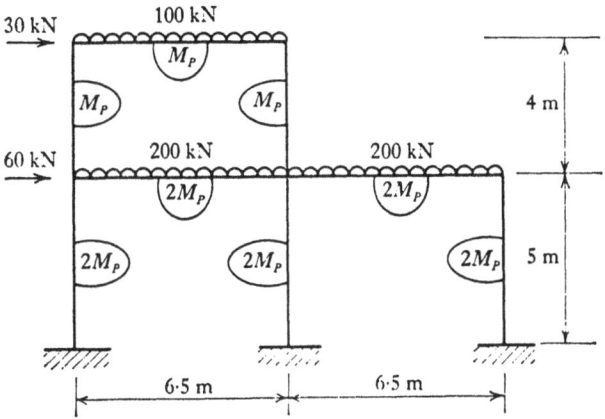

4.16 The rigid-jointed frame shown consists of prismatic members with the following full plastic moments:

$$AB, CDE, FG \qquad 150 \text{ kNm,}$$
$$BC, DF \qquad 240 \text{ kNm.}$$

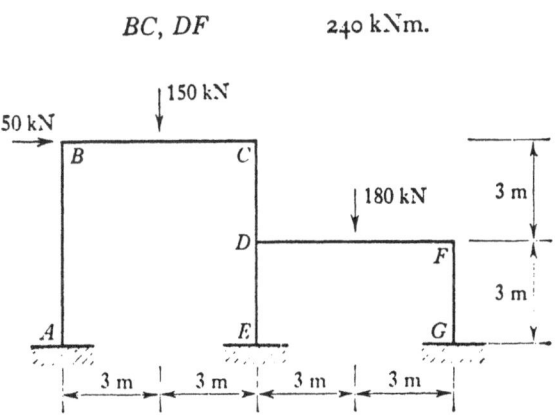

The frame carries the working loads indicated. Determine the load factor at collapse, ignoring any reduction in the full plastic moments due to axial load. It may also be assumed that the bending moments introduced by sway deflexions are negligible. (*M.S.T.* II, 1955; adapted.)   (*Ans.* 1·44.)

4. THE COMBINATION OF MECHANISMS

4.17 The rigidly jointed frame shown is composed of members with the following uniform full plastic moments:

$$AB = 150, \quad CD = 200, \quad AC = BD = 50, \quad CE = DF = 100 \text{ kNm.}$$

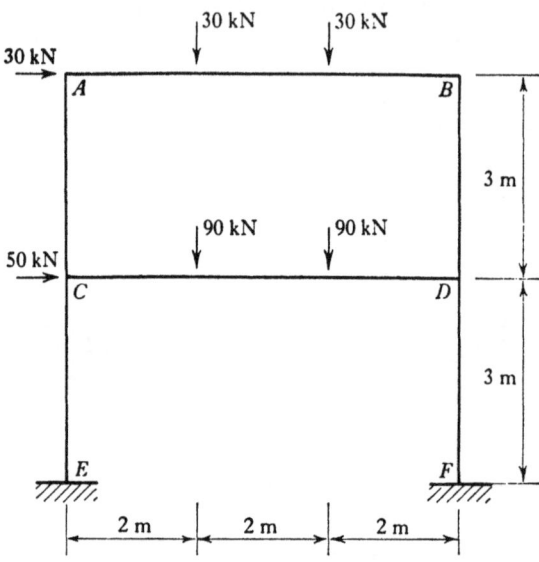

Determine the factor by which the loads shown would have to be multiplied for collapse just to occur. The effect of axial loads on the full plastic moments and the effect of deformation on the equations of equilibrium are to be neglected. (*M.S.T.* II, 1956; adapted.) (*Ans.* 1·375.)

4.18 The members of the rigidly jointed, fixed-base steel frame, whose dimensions and loading are as shown, are all uniform and have the same full plastic moment M_p. Making the usual assumptions of the simple plastic theory, find the value of W which would just cause plastic collapse. The effect of axial load on the value of the full plastic moment is to be ignored.

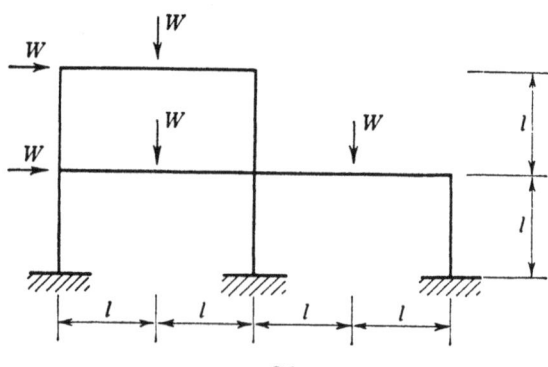

186

The solution will not be regarded as complete unless a bending-moment distribution throughout the frame is obtained for the condition of collapse such that the bending moment does not exceed M_p in magnitude at any section. (*M.S.T.* II, 1960.) (*Ans.* $2 \cdot 5 M_p / l$.)

4.19 The symmetrical two-bay pitched-roof frame shown may be considered encastré at the eaves A and B, and to have full-strength connexions between the uniform rafters. The valley C is supported by a simple pin-ended prop. Each rafter carries a point load at mid-span, as shown, the left-hand bay supporting loads W and the right-hand bay supporting loads kW.

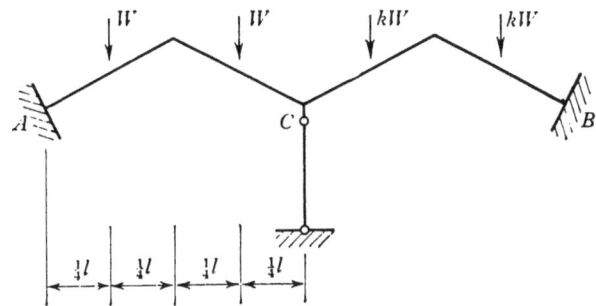

A plastic design is made for the frame, to a load factor $1 \cdot 75$, when the frame carries equal loads on both bays ($k = 1$). Show that, for this design, if k is reduced to $\frac{3}{4}$, the load factor drops to about $1 \cdot 62$, and that if k is further reduced to $\frac{1}{2}$, the load factor drops to about $1 \cdot 46$. (*M.S.T.* II, 1962.)

(May be simpler if the method of combination of mechanisms is not used.)

4. THE COMBINATION OF MECHANISMS

4.20 Find the collapse load factor for the frame shown.   (*Steel Skeleton*, II.)

(*Ans.* 2·31.)

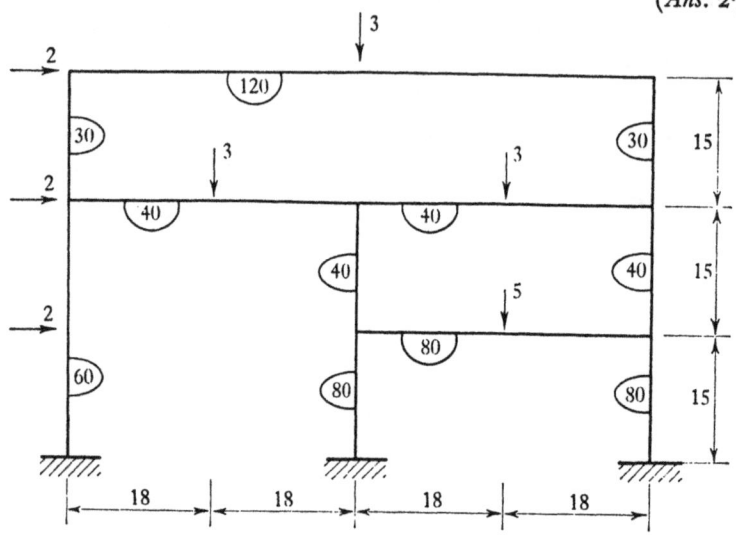

4.21 The rigid-jointed frame shown is composed of members of uniform plastic moment of resistance M_p. Find the value of W at which collapse of the frame shown in full lines would just occur. Determine the corresponding distribution of bending moment, and show it clearly on a diagram of the structure. Reduction of M_p due to the effects of axial load and shear may be neglected.

Hence or otherwise find the value of W at which collapse would occur in theory in a similar frame with a large number of bays (dashed lines). (*M.S.T.* II, 1964.)   (*Ans.* $2\cdot4M_p/L$, $2M_p/L$.)

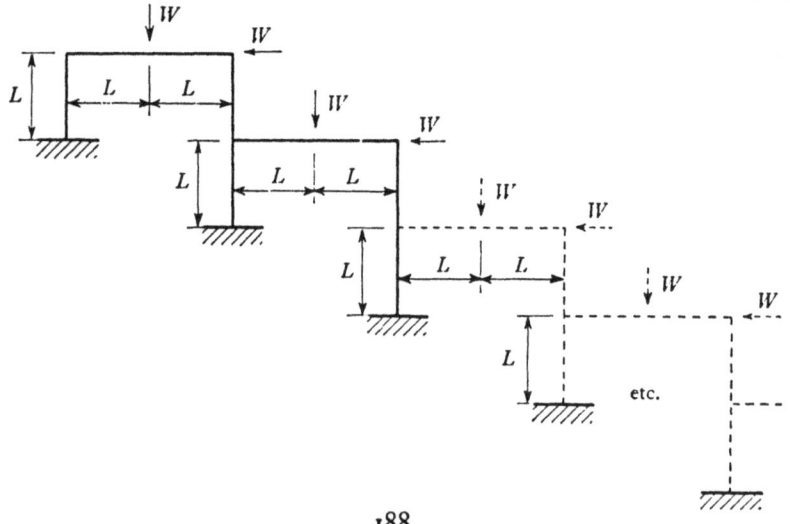

4.22 Repeat example 4.21 for the case of n bays. $\left(Ans. \dfrac{4(n+1)}{(2n+1)}.M_p/L.\right)$

4.23 The single-bay multi-storey frame shown consists of n storeys each of height l. The plastic moment of resistance of each beam is M_p and the columns are infinitely rigid. A load factor of 1·75 is required against collapse

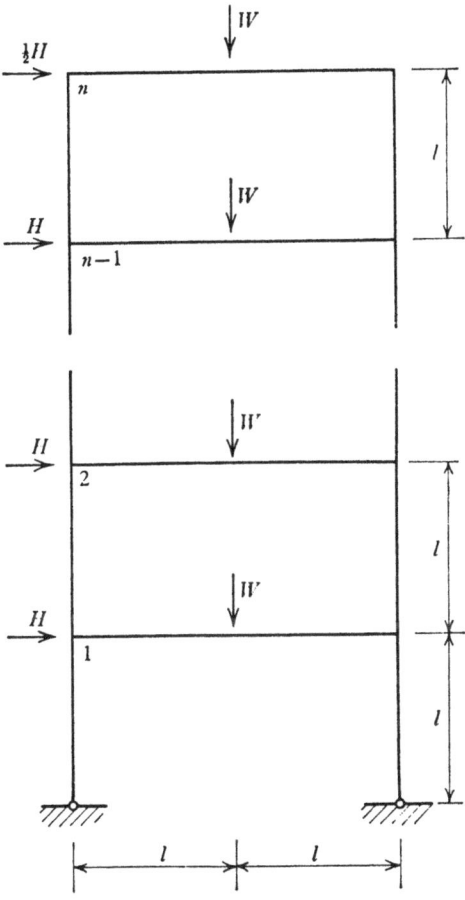

due to vertical loads only and a load factor of 1·4 is required against collapse involving wind loads. Determine the value of n in terms of W/H for all the possible modes of collapse of the frame. (Swansea University College, 1961.) (Do not use method of combination of mechanisms.)

(*Ans.* Beam mechanism when $n < W/2H$.
Pure sway mechanism when $n > 2W/H$.
Combined mechanism when $W/2H < n < 2W/H$.)

4. THE COMBINATION OF MECHANISMS

4.24 The figure shows a rigidly jointed, single bay, multi-storey frame in which all storey heights are equal, the stanchions being pin-connected to the foundations. The beams are all of the same uniform cross-section with a full plastic moment M_p. Each beam carries a uniformly distributed load W, while wind produces a concentrated load on the frame of H at each beam level.

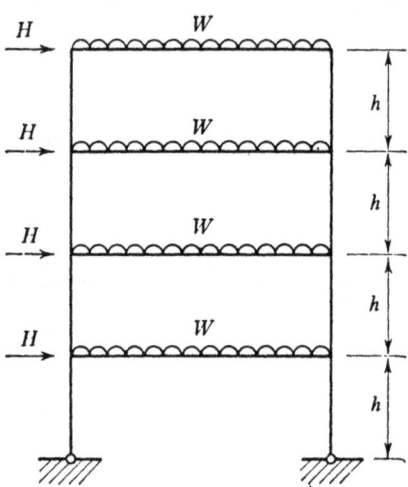

With loads W and H acting together, plastic collapse of the frame just occurs if $M_p = M_1$. If the loads W act alone, plastic collapse just occurs if $M_p = M_2$, while if the loads H act alone, plastic collapse just occurs if $M_p = M_3$. If in all cases plastic hinges occur only in the beams, show that, for a frame of *any number* of storeys, the following relations between M_1, M_2 and M_3 are satisfied:

$$M_1 = \frac{(4M_2 + M_3)^2}{16M_2} \quad \text{for} \quad M_3 \leqslant 4M_2;$$

$$M_1 = M_3 \quad \text{for} \quad M_3 \geqslant 4M_2.$$

The effect of elastic deformations on the collapse load and the effect of axial loads on the value of the full plastic moment are to be ignored. (*M.S.T.* II, 1960.)

5

EXAMPLES OF COMPLEX FRAMES

While the geometry of the rectangular frame makes it relatively simple to analyse, the plastic theory is, of course, applicable to many other forms of structure, such as grillages and space frames. Some simple examples of these and other more complex kinds of frame will be given, using straightforward methods of calculation. The same limitations on the use of plastic theory apply; that is, deflexions are assumed small, and instability is supposed to be prevented. Equally, if these assumptions are obeyed, then the basic theorems of plastic theory are valid. However, the theory so far presented will be in some cases insufficient for the full exploration of certain types of structure; this will be noted where relevant, and the topics will be treated further in vol. 2.

5.1 Grillages

The type of grillage to be considered consists of straight members all lying in the same plane, and subjected to loads acting perpendicular to the plane. Where the members cross they are taken to be joined with full strength connexions so that any required forces and moments can be transmitted through the grillage.

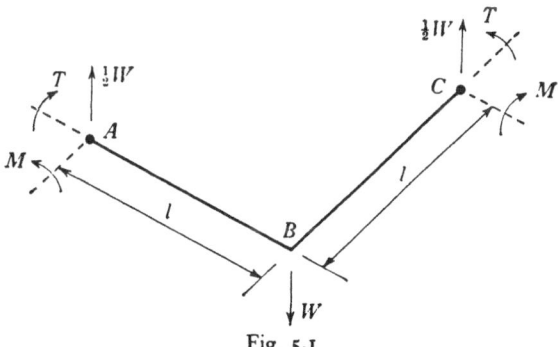

Fig. 5.1

A very simple example of such an elementary space frame is shown in fig. 5.1. The right-angled bent ABC lies in a horizontal plane, and is rigidly built-in to fixed walls at A and C. The bent carries a single vertical load at B. The collapse mechanism will evidently occur by the formation of plastic hinges at the ends A and C, as shown. Now any deformation of the bent will involve twisting of the members AB and BC, so that it is to be expected that, in addition to a bending moment M acting at the ends, there will also be a twisting moment T. It is the presence of combined bending and twisting that makes the plastic analysis (and, incidentally, the elastic analysis) of this structure difficult.

From the statics of fig. 5.1, it will be seen that

$$\tfrac{1}{2}Wl = M + T, \tag{5.1}$$

and no further information can be obtained from purely equilibrium considerations. However, experiment or theory will furnish a relation between M and T which governs the formation of a plastic hinge. Such a relationship will depend on the cross-section of the member, and it might well be of the form

$$\left(\frac{M}{M_p}\right)^2 + \left(\frac{T}{T_p}\right)^2 = 1, \tag{5.2}$$

where M_p is the value of the full plastic moment in the absence of twisting moment, and T_p might be called the full plastic torque in the absence of bending.

Although the mechanism condition is apparently satisfied by the arrangement of hinges in fig. 5.1, the equilibrium condition by the writing of equation (5.1), and the yield condition by the writing of equation (5.2) for the plastic hinges at A and C, there is not enough information for the solution of the problem. The twisting moment T may be eliminated from the two equations to give

$$\tfrac{1}{2}Wl = M + \frac{T_p}{M_p}\sqrt{(M_p^2 - M^2)}, \tag{5.3}$$

but the value of M is not yet determined. Full discussion of this type of problem is deferred to vol. 2; it turns out that the values of M and T at a hinge such as A or C in fig. 5.1 must be adjusted so that the plastic work done in any motion of the collapse mechanism is a *maximum*.

Thus, in equation (5.3), the value of M must be chosen to make the corresponding value of W a maximum, that is

$$1 - \frac{T_p}{M_p}(M_p^2 - M^2)^{-\frac{1}{2}}.M = 0,$$

or
$$M = \frac{M_p^2}{\sqrt{(M_p^2 + T_p^2)}}. \tag{5.4}$$

Substitution back into equation (5.3) gives

$$\tfrac{1}{2}W_c l = \sqrt{(M_p^2 + T_p^2)}. \tag{5.5}$$

This determination of the collapse load as a relative maximum for any chosen class of mechanism can be related to the normality condition (or flow rule) for hinges formed under the action of two (or more) parameters, as in equation (5.2). This condition relates possible deformations at hinges such as A or C in fig. 5.1 directly to the yield equation (5.2).

Even this trivial problem of the right-angled bent, therefore, raises difficulties if an exact solution is sought. If discussion is confined, however, to grillages composed of I-sections, then some simplifications may be made. For such sections, the strength in twisting is small; in equations such as (5.5), T_p^2 may be neglected compared with M_p^2, so that the collapse load is given by $\tfrac{1}{2}W_c l = M_p$. This simplification is equivalent to ignoring the twisting moment T in fig. 5.1, and the yield condition (5.2) reverts to the familiar $M^2 = M_p^2$. Indeed, it is really only for tubular or box sections that any appreciable twisting moment can be developed; further examples given here will assume that all twisting effects are negligible.

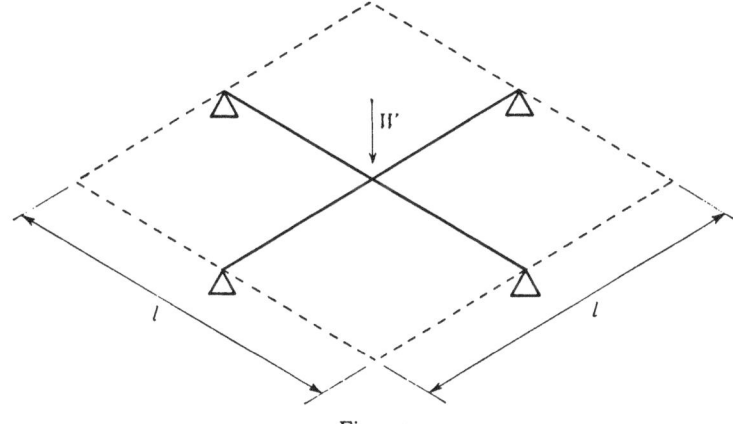

Fig. 5.2

The simple square grillage of fig. 5.2 consists of two equal simply-supported beams, each of span l and full plastic moment M_p; the two beams are connected rigidly at their centres, and a central concentrated load W is carried. (For this problem there is in fact no twisting of the members, due to symmetry.) Clearly the two beams act in series, and the collapse load is given by

$$\tfrac{1}{4}Wl = 2M_p. \tag{5.6}$$

This equation could have been derived by writing the work balance for a small movement δ of the load, fig. 5.3. The hinge rotation at the centre of each beam is $4\delta/l$, so that

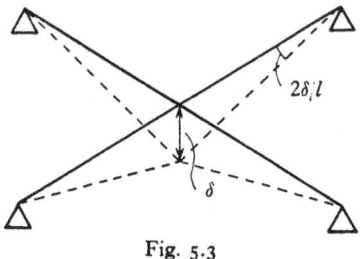

$$W\delta = M_p(4\delta/l) + M_p(4\delta/l), \quad (5.7)$$

or $W = 8M_p/l$ as before. From equation (5.7), it will be seen that had the two beams had unequal full plastic moments M_1 and M_2, then

Fig. 5.3

$$W = (M_1 + M_2)(4/l). \tag{5.8}$$

A grillage consisting of six beams, each of span $4l$ and full plastic moment M_p, is shown in fig. 5.4. Deformation of this grillage must involve twisting of the members, but the corresponding twisting

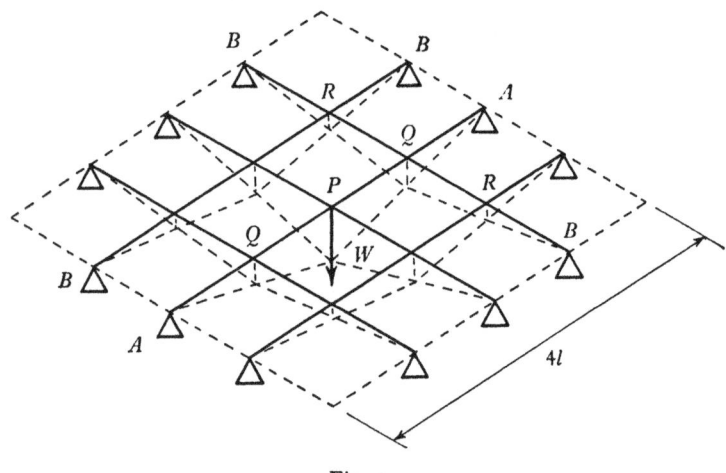

Fig. 5.4

moments will be ignored. If a single point load W acts at the centre of the grillage, then a possible collapse mechanism will result if a plastic hinge is developed at the centre of length of each of the six beams. This mechanism is indicated in fig. 5.4. Figure 5.5(a) shows the deflected form of each of the two central beams $AQPQA$; if the movement of the load point P is δ, then the deflexion of the point Q will be $\frac{1}{2}\delta$, and the central hinge rotation δ/l. Figure 5.5(b) shows the corresponding

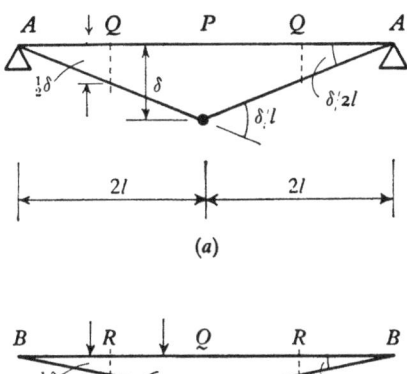

(a)

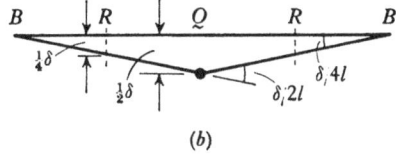

(b)

Fig 5.5

deflected form of the remaining four beams, with central hinge rotation $\frac{1}{2}\delta/l$. Thus the work equation is

$$W\delta = (2)M_p(\delta/l) + (4)M_p(\tfrac{1}{2}\delta/l),$$

or
$$W = 4M_p/l. \qquad (5.9)$$

This value of W must be regarded as an upper bound, pending a statical analysis. The grillage is statically determinate at collapse, and forces and moments are shown in fig. 5.6; corresponding bending-moment diagrams are displayed in fig. 5.7. It will be seen that no bending moment in the grillage exceeds the value M_p, so that equation (5.9) gives the correct value of the collapse load.

The reader may wish to check that the collapse load of the same grillage, acted upon by a load W at *each* of the nodes (i.e. a total load of $9W$), is given by $W = M_p/l$, and that the collapse mechanism is unchanged.

For the grillage shown in plan in fig. 5.8, the ends of the beams are pinned to abutments which can resist both upward and downward forces. The beam of length $4l$ has a uniform full plastic moment M_1, and the two short beams have full plastic moment M_2. The grillage is to be designed to carry a single point load $4W$ placed anywhere on the longer beam. It will be assumed first that the value of M_2 is such that the short beams do not participate in the collapse mechanism; the corre-

5. EXAMPLES OF COMPLEX FRAMES

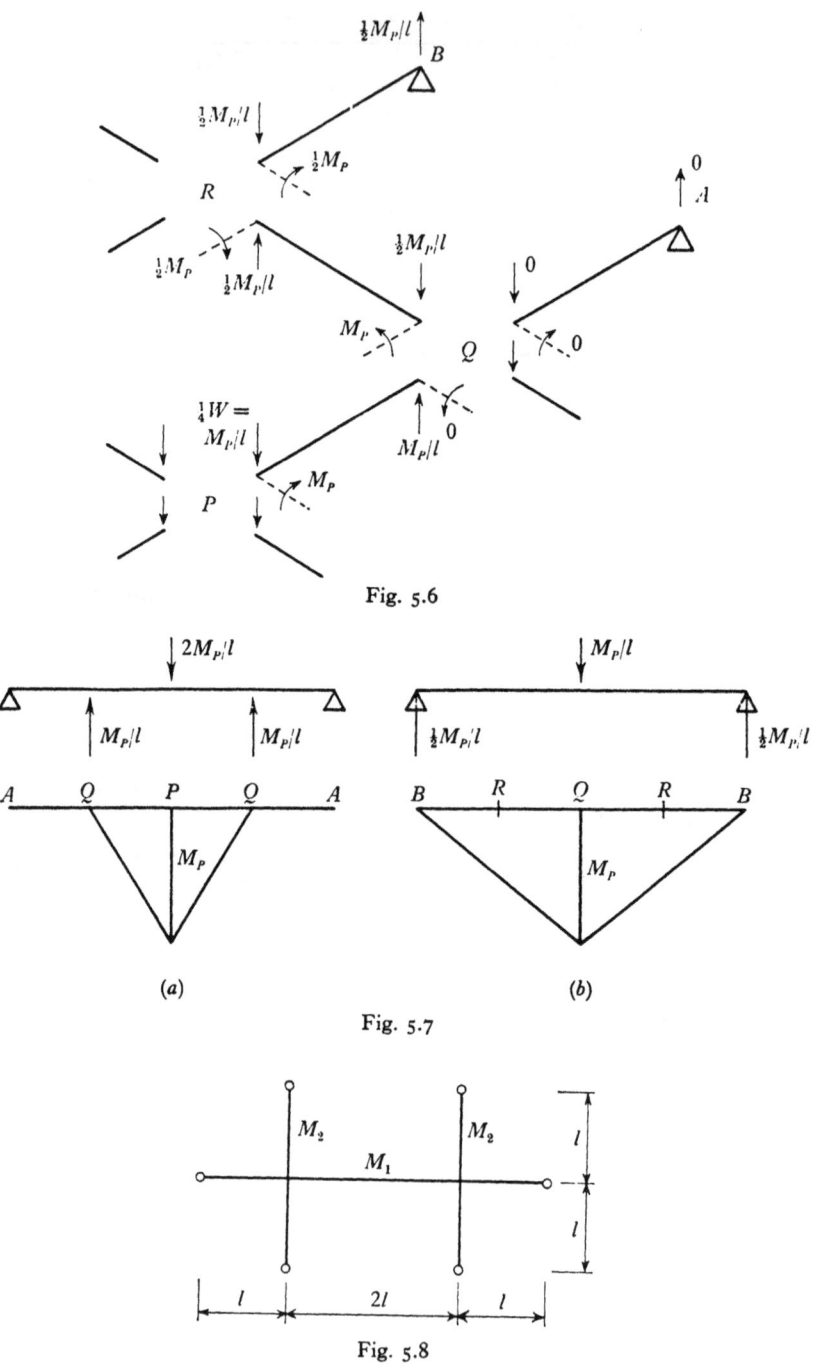

Fig. 5.6

(a)

(b)

Fig. 5.7

Fig. 5.8

196

sponding value of M_2 will be found later. The longer beam can then be considered as a continuous beam on four supports, fig. 5.9(a), and it is quickly evident that the most unfavourable position of the point load is at the centre of span as shown. From the bending-moment diagram of fig. 5.9(b), the value of M_1 is given by

$$M_1 = Wl. \qquad (5.10)$$

Corresponding to this collapse bending-moment diagram, the abutments must provide downward reactions of value W, so that the load transmitted to each of the shorter beams is $3W$.

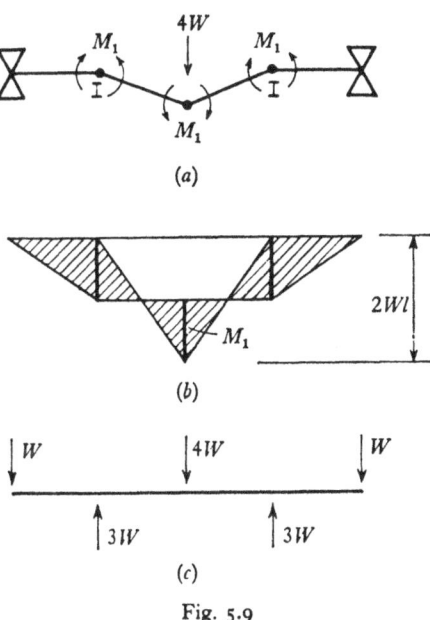

(a)

(b)

(c)

Fig. 5.9

The bending moment at the centre of each of the shorter beams therefore has value $\frac{3}{2}Wl$, so that a possible design would be

$$M_2 = \tfrac{3}{2}Wl = \tfrac{3}{2}M_1. \qquad (5.11)$$

However, the design given by equations (5.10) and (5.11) is not satisfactory when the load $4W$ moves away from the centre of the grillage. Despite the fact that the worst position of the load for the design of the long beam is at centre span, there is a more critical position for the grillage as a whole.

5. EXAMPLES OF COMPLEX FRAMES

To see this, the collapse of the grillage having $M_1 = Wl$, $M_2 = \frac{3}{2}Wl$ will be investigated. A possible collapse mechanism (which is to be confirmed) is shown in fig. 5.10(a), involving the collapse of one of the shorter beams; the load factor on the load $4W$ is shown as λ, and the value of the parameter z defining the position of the sagging hinge in the longer beam is as yet unknown.

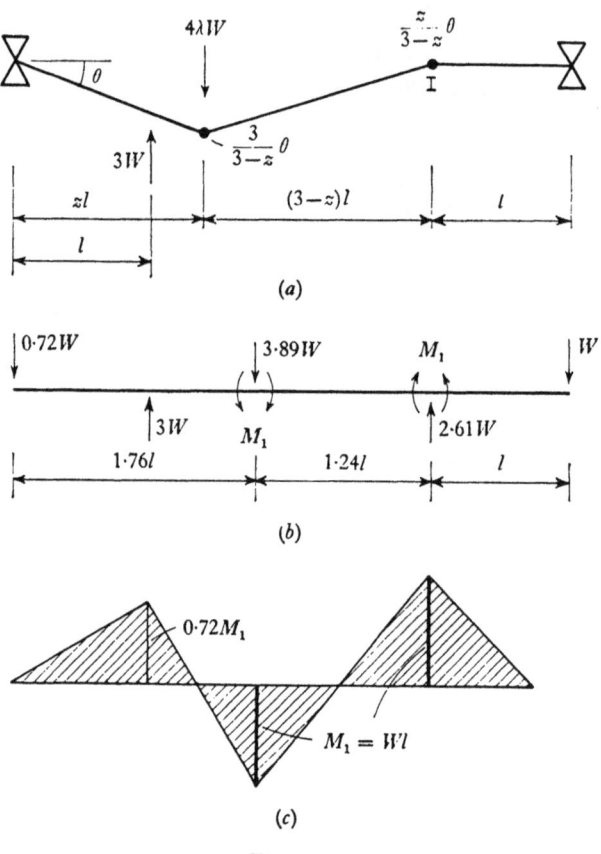

Fig. 5.10

Since the shorter beam is collapsing, it must be subjected to a load $3W$ for which it was designed (fig. 5.9(c) and equation (5.11)). The work equation for the mechanism of fig. 5.10(a) is therefore

$$4\lambda W(zl\theta) - 3W(l\theta) = M_1\left(\frac{3+z}{3-z}\right)\theta = Wl\left(\frac{3+z}{3-z}\right)\theta,$$

198

i.e.
$$\lambda = \frac{1}{2}\left[\frac{6-z}{z(3-z)}\right]. \tag{5.12}$$

The load factor λ is a minimum for $z = 6 - 3\sqrt{2} = 1\cdot757$, and the corresponding value of λ is $\frac{1}{8}(3 + 2\sqrt{2}) = 0\cdot971$.

The forces acting on the beam under these conditions are shown in fig. 5.10(b), and the bending-moment diagram in fig. 5.10(c). It will be seen (i) that the bending moments in the longer beam satisfy the yield condition, and (ii) that the load on the short beam which is not supposed to collapse has value $2\cdot61W$, which is in fact less than the collapse load of $3W$. Thus the correct solution has been found, and the conclusion must be drawn that the grillage designed to have the full plastic moments of (5.10) and (5.11) would not carry the point load of $4W$ traversing (slowly) the long beam. The load factor under these conditions is $0\cdot97$, so that if both M_1 and M_2 were increased by 3%, the design would be satisfactory. Alternatively, the reader may wish to check that a satisfactory design would also result if M_1 were left unchanged at the value Wl, and M_2 were increased by some 6% to the value $1\cdot60Wl$.

5.2 A simple space frame

The plane frame of fig. 5.11(a) is composed of four equal members, each of full plastic moment M_p, joined together rigidly at the corners. When subjected to equal and opposite diagonal loads W, the collapse mechanism will be that sketched in fig. 5.11(b), and

$$(2)W\left(\frac{l}{\sqrt{2}}\right)\theta = M_p(8\theta),$$

or
$$Wl = 4\sqrt{2}M_p. \tag{5.13}$$

The same square frame is mounted horizontally on four equal tubular legs, each of full plastic moment M_p and length $l/6$, as shown in fig. 5.12. Each leg is pinned at one end with a universal joint to the frame, and fixed at the other end in the ground, so that the maximum load that can be carried by each leg, acting as a cantilever, is $6M_p/l$. The frame is subjected to a single diagonal load W, and the problem is to determine the collapse value of this load.

Supposing first that no plastic hinges are formed within the square frame itself, then a possible collapse mode is that shown in fig. 5.13(a), in which each of the cantilever legs forms a hinge at its foot. Since each

5. EXAMPLES OF COMPLEX FRAMES

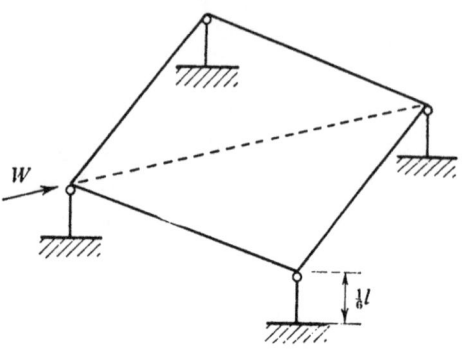

Fig. 5.12

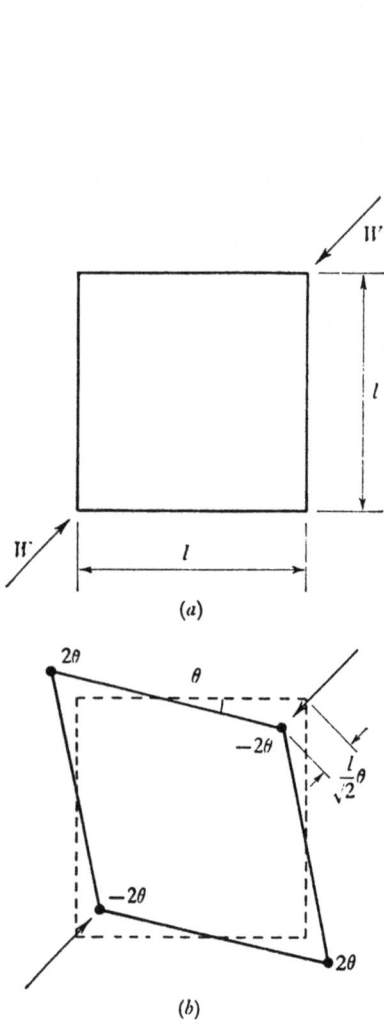

(a)

(b)

Fig. 5.11

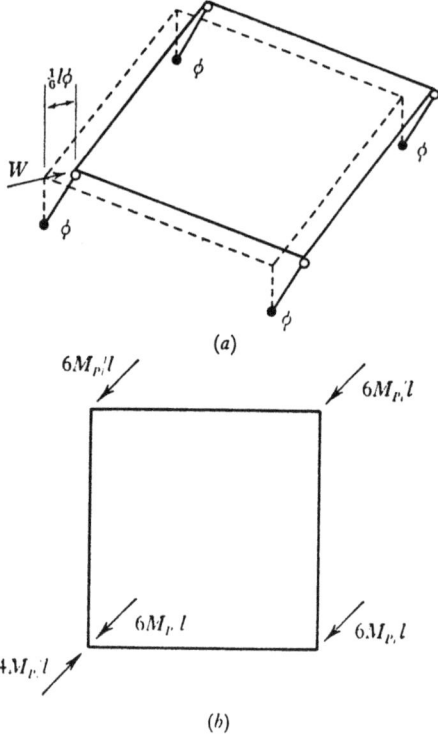

(a)

(b)

Fig. 5.13

leg requires a load $6M_p/l$ for its collapse, the value of W is $24M_p/l$; this can also be seen from the work equation:

$$W\left(\frac{l}{6}\phi\right) = M_p(4\phi),$$

or
$$W = 24M_p/l. \tag{5.14}$$

As usual, this value of W must be regarded as an upper bound pending a satisfactory static check.

The square frame is acted upon by the planar forces of fig. 5.13(b), if collapse is occurring by the mode of fig. 5.13(a). The question to be answered is whether or not the square frame can carry these forces, that is, whether a bending-moment distribution can be constructed which does not violate the yield condition. This in turn resolves itself into the question of determining values of M_1, M_2 and M_3 in fig. 5.14 such that all three are less than M_p.

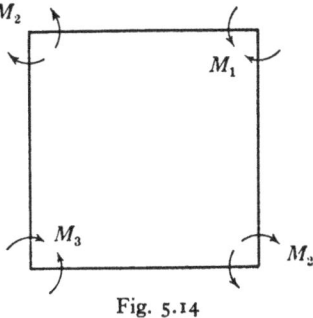

Fig. 5.14

Now the 'ring' frame of fig. 5.14 has, basically, three redundancies, but one of these should be determinable from considerations of symmetry. The easiest way of obtaining a relationship between the three bending moments of fig. 5.14 (which are in equilibrium with the forces of fig. 5.13(b)), is to consider a virtual deformation of the frame. Figure 5.11(b) is a suitable *virtual* mechanism, which leads directly to the equation

$$\left(\frac{24M_p}{l} - \frac{6M_p}{l}\right)\left(\frac{l}{\sqrt{2}}\theta\right) + \left(\frac{6M_p}{l}\right)\left(\frac{l}{\sqrt{2}}\theta\right)$$
$$= (-M_1)(-2\theta) + (M_2)(4\theta) + (-M_3)(-2\theta),$$

or
$$6\sqrt{2}M_p = M_1 + 2M_2 + M_3. \tag{5.15}$$

Thus, if M_1 and M_2 were known, M_3 would be calculable; for example, if M_1 and M_2 had their maximum permitted values of M_p, then

$$M_3 = (6\sqrt{2}-3)M_p.$$

This calculation shows at once that it is not possible to find any values of M_1, M_2 and M_3 such that all three are less than M_p. Indeed, the 'best' values would be given from equation (5.15) as

$$M_1 = M_2 = M_3 = \tfrac{3}{2}\sqrt{2}M_p. \tag{5.16}$$

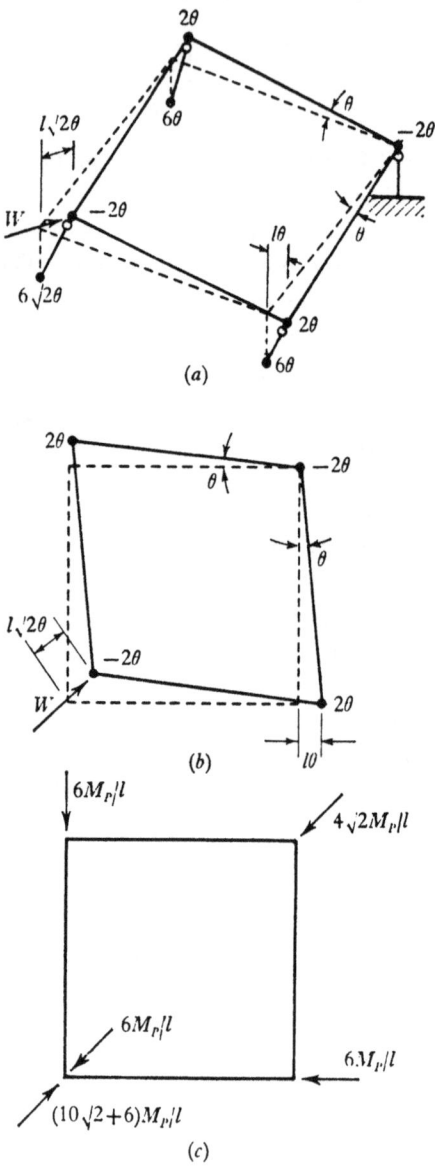

Fig. 5.15

However, as usual, a lower bound on the collapse load may be constructed from (5.16); if all values in fig. 5.13 were multiplied by the factor $2/3\sqrt{2}$, then it would be just possible to construct a satisfactory bending-moment diagram for the frame with $M_1 = M_2 = M_3 = M_p$. Thus, using (5.14),

$$\frac{24M_p}{l} \geqslant W_c \geqslant 8\sqrt{2}\,\frac{M_p}{l}. \tag{5.17}$$

The correct collapse mode is not immediately obvious, but may be determined to be that shown in perspective in fig. 5.15 (a) and in plan in fig. 5.15 (b); careful note should be taken of the various hinge rotations marked in the figures. The work equation gives

$$W(l\sqrt{2}\theta) = M_p[8\theta + 6\sqrt{2}\theta + 2(6\theta)],$$

or
$$W = (10\sqrt{2} + 6)M_p/l. \tag{5.18}$$

This value of W ($20 \cdot 14 M_p/l$) is less than the upper bound of (5.17); a statical analysis must be made in order to check that the bending moment at the foot of the leeward leg is less than M_p. The planar forces are shown in fig. 5.15 (c), and it will be seen that a total force of $4\sqrt{2}M_p/l$ must be developed at the top of the leeward leg; this, however, is less than the permitted value of $6M_p/l$, so that the solution is completely satisfactory.

The dimensions of the frame in this example have been chosen with some care, in order to make it relatively simple. Had the legs in fig. 5.12 been longer, say $l/4$ instead of $l/6$, then the problem becomes very much more difficult. Following the previous order of working, the first trial collapse mode of fig. 5.13 (a) would give, for the taller frame (cf. (5.14)),

$$W(\tfrac{1}{4}l\phi) = M_p(4\phi),$$

or
$$W = 16M_p/l. \tag{5.19}$$

The virtual work analysis of equation (5.15) then gives the equation

$$4\sqrt{2}M_p = M_1 + 2M_2 + M_3, \tag{5.20}$$

and it will be seen that no values of the bending moments M_1, M_2 and M_3 can be found to satisfy this equation and at the same time remain below the fully plastic value. The 'best' solution is

$$M_1 = M_2 = M_3 = \sqrt{2}M_p,$$

so that, from (5.19),

$$\frac{16M_p}{l} \geqslant W_c \geqslant 8\sqrt{2}\,\frac{M_p}{l}. \tag{5.21}$$

On the other hand, the collapse mode of fig. 5.15(*a*) and (*b*) is also unsatisfactory. The work equation for this mode (cf. (5.18)) gives

$$W(l\sqrt{2}\theta) = M_p[8\theta + 4\sqrt{2}\theta + 2(4\theta)],$$

or
$$W = (8\sqrt{2}+4)M_p/l. \tag{5.22}$$

The corresponding planar forces are shown in fig. 5.16, and it will be seen that a force of $4\sqrt{2}M_p/l$ is required at the leeward leg, whereas a leg of length $l/4$ can carry a force of only $4M_p/l$.

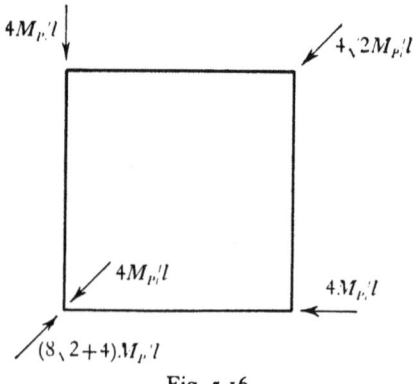

Fig. 5.16

The correct collapse mode is in this case a combination of the modes of fig. 5.13(*a*) and fig. 5.15(*a*), which leads apparently to a mechanism of *two* degrees of freedom, fig. 5.17(*a*). However, it is found that there is sufficient information for the problem to be determined uniquely.

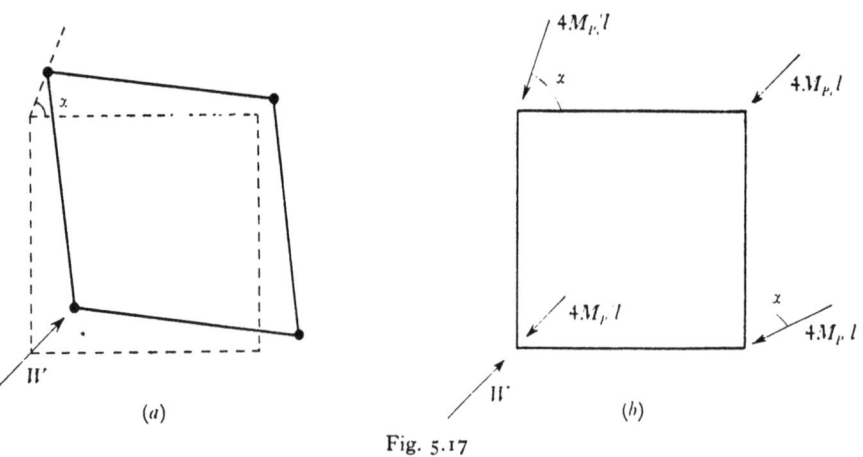

(*a*)   (*b*)

Fig. 5.17

The planar forces acting on the square frame are shown in fig. 5.17(b); each leg, since it is collapsing, applies a force of magnitude $4M_p/l$. The unknown angle α is marked in fig. 5.17; in fact, α could be determined in terms of θ and ϕ in figs 5.15(a) and 5.13(a), but this is not necessary. Two relationships may be found between the forces of fig. 5.17(b), as follows. The *virtual* mechanism of fig. 5.13(a), in which the corners of the square frame all move through the same distance (say d) in the direction of the load W, leads to the equation

$$\left(W - \frac{4M_p}{l}\right)d - 2\left(\frac{4M_p}{l}\right)d\cos(\alpha - 45°) - \left(\frac{4M_p}{l}\right)(d) = 0,$$

or
$$W = \frac{8M_p}{l}\left(1 + \frac{1}{\sqrt{2}}\cos\alpha + \frac{1}{\sqrt{2}}\sin\alpha\right); \qquad (5.23)$$

this equation, of course, does no more than express horizontal equilibrium.

The *virtual* mechanism of fig. 5.15(b), taken with the forces of fig. 5.17(b), gives

$$\left(W - \frac{4M_p}{l}\right)(l\sqrt{2}\,\theta) - 2\left(\frac{4M_p}{l}\sin\alpha\right)(l\theta) = 8M_p\theta,$$

where the term on the right-hand side is the virtual work corresponding to the plastic hinges at the corners of the square frame; this equation reduces to
$$W = \frac{8M_p}{l}\left(\frac{1}{2} + \frac{1}{\sqrt{2}} + \frac{1}{\sqrt{2}}\sin\alpha\right). \qquad (5.24)$$

Equations (5.23) and (5.24) may be solved simultaneously to give

$$\left. \begin{array}{c} \cos\alpha = 1 - \dfrac{1}{\sqrt{2}}, \quad \alpha = 72° \; 58', \\[2mm] W = 15{\cdot}07M_p/l. \end{array} \right\} \qquad (5.25)$$

Applying again the uniqueness theorem, it is quite certain that the value given in (5.25) is the correct value of the collapse load, despite the curious mechanism of collapse. However it *is* a mechanism, and both the yield condition and equilibrium are satisfied.

Tubular legs were assumed for the frame, so that the full plastic moment was the same for any axis of bending. Had a rectangular or I-section been used, then the plastic hinges would have formed under a combination of two bending moments, one acting about the strong axis of the section and one about the weak axis. Such a combination of two bending moments is analogous to the formation of a plastic hinge under bending and twisting, and leads in fact to the same sorts of difficulty that were met with in the grillage problem. Thus the mechanism of fig. 5.17(a) only *appears* to have two degrees of freedom; the angle α is determinable, as was seen, and this could be regarded as another application of the concept of maximum plastic work, which is not dealt with in this volume. For this reason, no further discussion will be given here of grillages or space frames.

5.3 The Vierendeel girder

The action of the Vierendeel girder is to resist external loading mainly by bending of the members. It should, therefore, be suitable for plastic analysis, and two brief examples will illustrate that the techniques already established can be used for this type of frame. As usual, it will be assumed that axial loads in the members do not cause instability; the effect on the values of the full plastic moments can be allowed for if necessary.

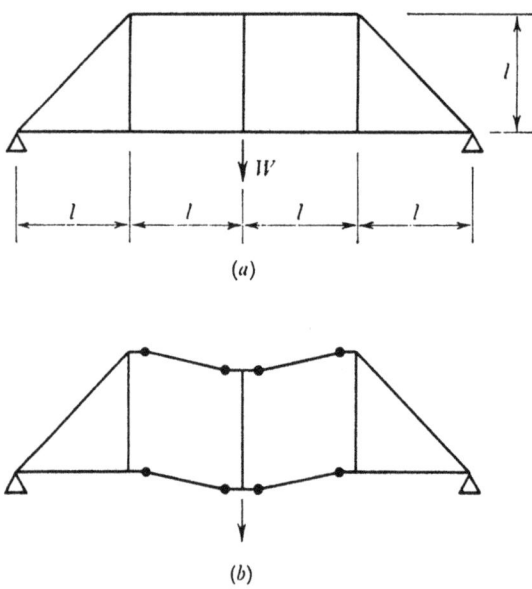

(a)

(b)

Fig. 5.18

The simply-supported girder of fig. 5.18(a) has rigid joints, and all members have the same full plastic moment M_p. The two end bays of the girder are triangulated, as will be seen, and cannot participate in any collapse mechanism (except possibly by rotation as rigid bodies). The collapse mechanism is therefore very simple, fig. 5.18(b), and the collapse equation is

$$Wl = 8M_p. \tag{5.26}$$

A slightly more complex girder is shown in fig. 5.19(a), and again all members will be taken to have the same value of full plastic moment, M_p.

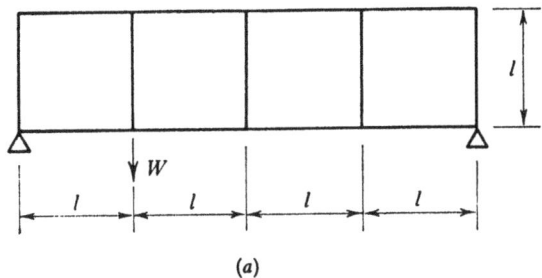

(a)

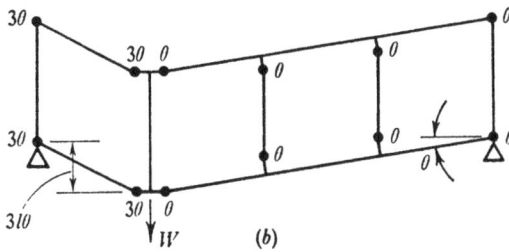

(b)

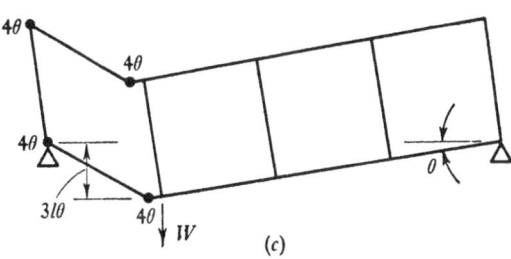

(c)

Fig. 5.19

As a first trial, the mechanism of fig. 5.19(b) will be assumed; this mechanism may not (in fact, does not) correspond to any possible equilibrium state, but it is clearly a possible mode of deformation, which will therefore lead to an upper bound for the collapse load. The collapse equation is

$$3Wl = 20M_p \tag{5.27}$$

so that
$$W_c \leqslant \frac{20}{3}\frac{M_p}{l}.\qquad(5.28)$$

It is not immediately obvious that the mechanism of fig. 5.19(b) represents an impossible equilibrium state, but this can be demonstrated in several ways. For example, equilibrium of the whole girder requires that the reaction at the left-hand support should be $\frac{3}{4}W$, that is $5M_p/l$ from equation (5.27). However, considering the left-hand bay to be a simple square portal frame on its side, it will be seen that the reaction at the left-hand support should be $4M_p/l$.

The correct collapse mechanism is shown in fig. 5.19(c), from which

$$3Wl = 16M_p,$$

or
$$W_c \leqslant \frac{16}{3}\frac{M_p}{l}.\qquad(5.29)$$

The value of the reaction at the left-hand support is now correct, and it is relatively easy to construct an equilibrium set of bending moments for the three right-hand panels that nowhere violates the yield criterion.

5.4 The arch

The assumptions of simple plastic theory, which have been noted in the previous pages and which will be commented upon more fully in chapter 6, are thrown into prominence by the problem of the arch, particularly when the rise is small. First, axial loads in the members of the arch will tend to be high. Allowance can be made for the effect on the value of the full plastic moment, but the problems associated with instability become more critical.

Second, and of great importance for shallow arches, the line of thrust will lie close to the arch rib. This means that any small deflexion of the arch, ignored in all simple plastic calculations, will have a more or less marked effect on the values of the bending moments; the change of geometry of the arch due to deflexions will upset the equilibrium equations.

Providing due care is taken, however, and for arches which are not too shallow, plastic theory can be used to give a good estimate of strength, and the single-span arch has many of the characteristics of the simple portal frame. For example, four hinges must, in general, be formed at collapse; the pinned-base arch of fig. 5.20(a) carrying a single load W will collapse as shown in fig. 5.20(b), two of the hinges in this case being those at the abutments.

The location of the hogging plastic hinge can be found graphically by means of the thrust line, as sketched in fig. 5.20(*a*). The basic property of the arch holds whether it is elastic or plastic: the *vertical*

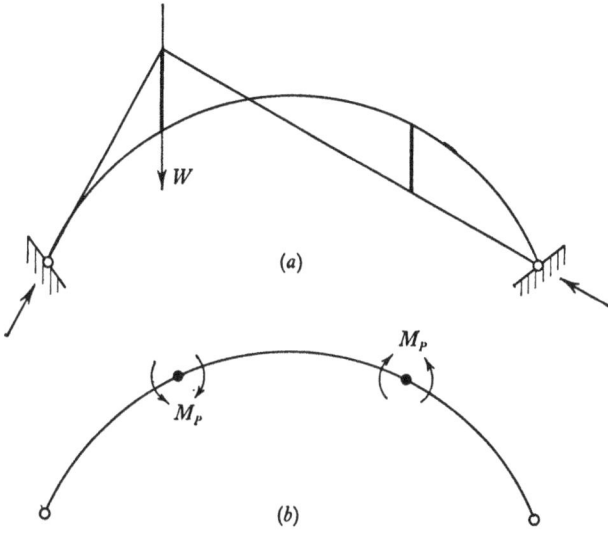

Fig. 5.20

intercept between the thrust line and the centre line of the arch, multiplied by the *horizontal* component of the abutment thrust, gives the value of the bending moment at any section. Thus the two intercepts in fig. 5.20(*a*) must be made equal (for an arch of constant full plastic moment M_p). Similarly, the same arch with fixed abutments will have a thrust line at collapse as shown in fig. 5.21.

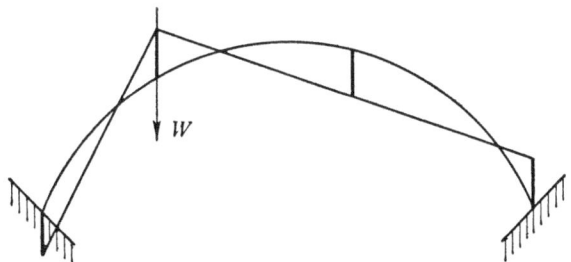

Fig. 5.21

5. EXAMPLES OF COMPLEX FRAMES

As a numerical example, the pinned-base arch composed of three uniform straight members, fig. 5.22, will be investigated. For the single load W acting at one of the knees, the thrust line must be positioned to give equal intercepts h_1; by simple geometry, $h_1 = \frac{1}{2}h$. Thus the value

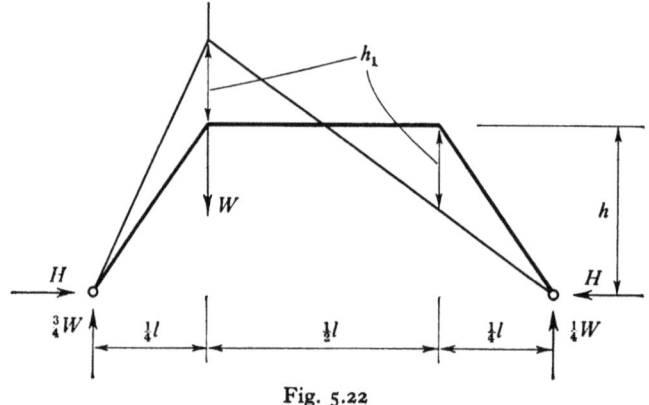

Fig. 5.22

of the horizontal abutment thrust H may be found by considering the triangle of forces at the right-hand abutment:

$$\frac{H}{\frac{1}{4}W} = \frac{\frac{1}{4}l}{\frac{1}{2}h},$$

or
$$Hh = \frac{1}{8}Wl. \tag{5.30}$$

Thus, if M_p is the full plastic moment of the members of the arch,

$$M_p = Hh_1 = \frac{1}{2}Hh = \frac{1}{16}Wl. \tag{5.31}$$

This same collapse relationship could also have been found by writing the work equation. Once the hinges have been located by sketching the thrust line, the collapse mechanism may be constructed. Thus, from fig. 5.23,
$$W(\frac{1}{4}l\theta) = M_p(4\theta), \tag{5.32}$$

which leads to the same result as before.

Similarly, the mechanism of fig. 5.23 for the *fixed*-base arch leads to the collapse equation
$$\frac{1}{4}Wl = 6M_p. \tag{5.33}$$

The use of a thrust line in the analysis of arches is analogous to the use of free and reactant bending-moment diagrams for beams and frames. Once again, therefore, there is a choice between working from the

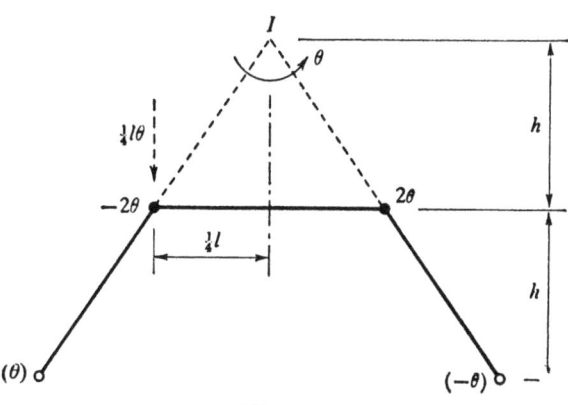

Fig. 5.23

principles of statics in constructing a thrust line, or using the work equation associated with a mechanism of collapse. Very often a mixture of the two approaches will give the deepest insight into structural behaviour, and will lead to the quickest calculations.

EXAMPLES

5.1 A simple plastic design is required for the square frame $ABCD$ under the loading shown; the value of W may be taken to incorporate a suitable load factor. The full plastic moment of member AB has value M_1, that of members BC, CD and DA has value M_2, and joints are full-strength.

Make three designs of the frame:

(i)   with M_1 as small as possible;
(ii) with $M_1 = M_2$; and
(iii) with M_2 as small as possible.   (*M.S.T.* II, 1968.)

(*Ans.* $M_1 = \frac{1}{16}Wl$, $M_2 = \frac{1}{4}Wl$; $M_1 = M_2 = \frac{1}{2}Wl(2-\sqrt{3})$; $M_1 = \frac{5}{32}Wl$, $M_2 = \frac{1}{8}Wl$.)

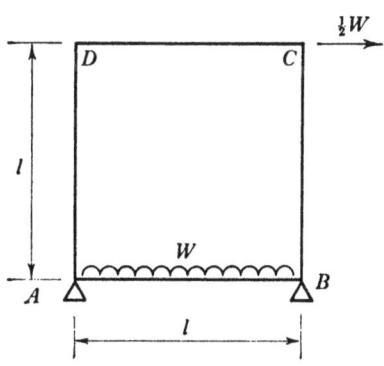

5. EXAMPLES OF COMPLEX FRAMES

5.2 The figure shows the plan of part of a jetty, protected by two continuous fenders PQ and RS. The fenders are connected to each other and to the jetty at equal distances, the short struts transmitting axial forces but not bending

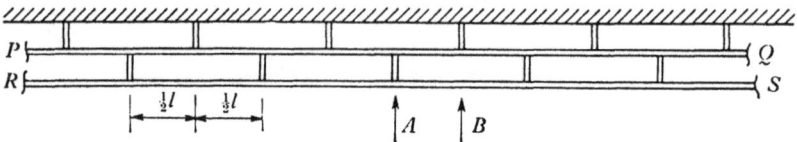

moments. Fenders PQ and RS have full plastic moments M_1 and M_2 respectively. Find the ratio M_2/M_1 so that the collapse loads for impact at A and B are the same. (*Ans.* 2.)

5.3 A simply-supported beam of span a and full plastic moment M_1 rests on two simply-supported beams of span b and full plastic moment M_2 as shown. If a point load W is to be placed anywhere along the beam of span a, find the

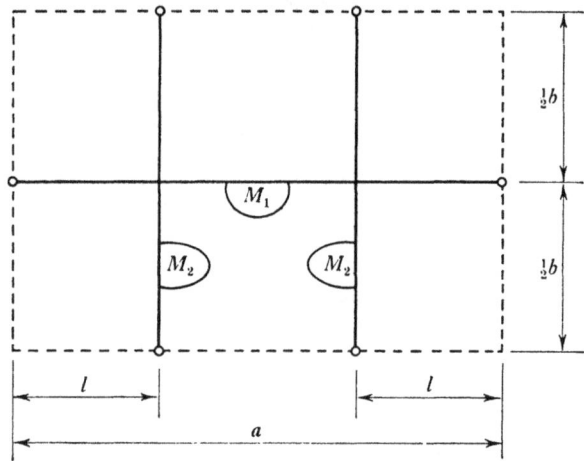

minimum possible value of M_1 and the corresponding value of l. For these values, find the smallest value of M_2. (Assume the beams are pinned rigidly to the abutments and hence cannot lift.)

(*Ans.* $0{\cdot}051\,Wa$, $l = 0{\cdot}296a$ [$= a/(26-26\sqrt{2})$], $0{\cdot}185\,Wb$.)

5.4 The Vierendeel girder shown has full-strength joints; the horizontal members have full plastic moment M_p and the vertical members have full plastic moment kM_p. Determine the value of the collapse load W, acting as shown, in terms of k, M_p and l, for all values of k. The usual assumptions of simple plastic theory may be made. (*M.S.T.* II, 1966.)

$$(Ans.\ k \leqslant 1,\ W = 3(1+k)M_p/l,\ k \geqslant 1,\ W = 6M_p/l.)$$

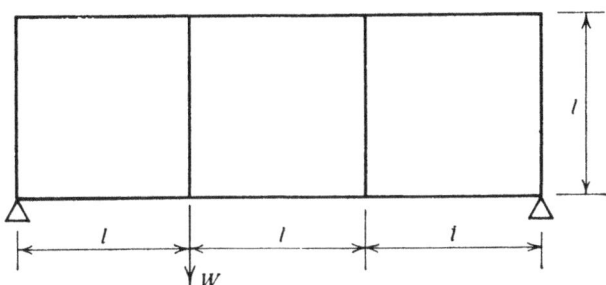

5.5 In the Vierendeel girder of fig. 5.18(a), the point load W is now applied at the first panel point (distant l from the support) rather than at mid-span. Show that the collapse value of W is $12M_p/l$, where M_p is the value of the full plastic moment of all members.

5.6 The symmetrical four-panel Vierendeel girder is to be designed with prismatic steel members by means of the plastic collapse method, with a load factor of 2·0. Determine suitable values for the plastic moment of resistance of each member assuming that no allowances are required for direct forces and any tendency towards instability.

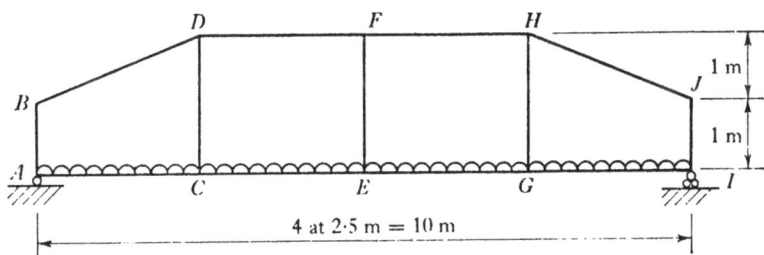

Determine the direct and shearing forces corresponding to your solution and briefly explain how modification to the M_p values could be made to allow for these effects. (Oxford University, Final Examination in Engineering Science, 1966; adapted.)

5.7 A symmetrical arch of the dimensions shown is composed of three straight members of uniform section of full plastic moment M_p. The connexions between the members are full strength, and the arch is pinned to rigid abutments.

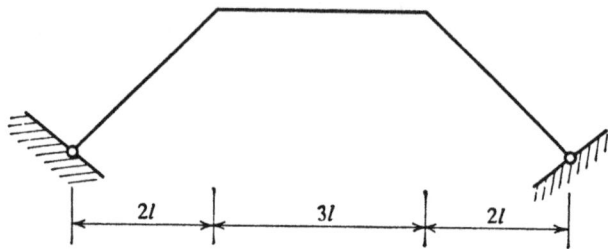

The arch is to be designed to resist plastic collapse under the action of a single concentrated vertical load W, wherever this load is placed on the arch. It may be assumed that the value of W incorporates a suitable load factor. Show that the minimum required value of M_p is, very closely, $0.5Wl$.

The arch members may be assumed to remain stable, and the effects of shear force and direct thrust on the value of M_p may be neglected. (*M.S.T.* II, 1961.)

5.8 The mild steel fixed-ended arch shown has constant cross-section throughout. Determine the value of P that will just cause collapse if the full plastic moment of the section is 2200 kNm. (Manchester University, Final Examination B.Sc. Tech. 1964; adapted.) (*Ans.* 2000 kN.)

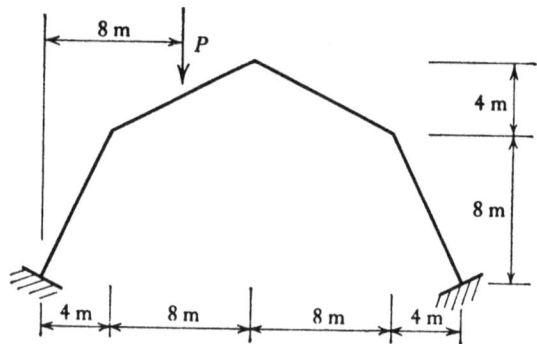

6

ASSUMPTIONS AND AIMS OF
PLASTIC DESIGN

Plastic theory has been presented here mainly with reference to *framed* structures, in which the action of external loads is resisted primarily by bending of the members. The theory can be applied successfully to other types of structure; this should be borne in mind, although these concluding remarks will continue to refer mainly to beams and frames.

The main object of plastic methods is to calculate the collapse loads of structures. Thus, of a large number of possible design criteria, plastic theory concentrates on one only, that of *strength*. In its simple form, the theory makes no attempt to assess deflexions, nor to enquire into the stability of individual members or of the structure as a whole. Thus, if he is to use plastic design methods, the designer must be satisfied that the particular structure under consideration is one for which strength is the overriding design criterion. Having designed the structure on the basis of plastic theory, the designer may, of course, estimate deflexions under working load, check the stability of members, and make any other calculations that seem to him to be desirable.

There is, however, rather more to the matter than a general and perhaps vague feeling on the designer's part that his structure should be suitable for plastic design. Firm reliance can only be placed on the plastic calculations if deflexions are small in a very definite technical sense and if the danger of instability is negligible. The proofs of the plastic theorems, and in particular of the safe theorem, depend on the ability of the members in a frame to form plastic hinges, which can undergo very large rotations without fall-off in the value of the full plastic moment. Any behaviour which interferes with these abilities will render the proofs of the theorems invalid, and must in turn lead to the structure as a whole being regarded with suspicion.

This situation is not new, nor is it confined to plastic calculations; it is merely that plastic theory enables a rather precise assessment to be made of required structural behaviour. For example, the elastic designer of a steel frame will usually, and unthinkingly, assume that deflexions

6. ASSUMPTIONS AND AIMS OF PLASTIC DESIGN

are small, in the sense that, in all calculations, original undeformed dimensions will be used in the equilibrium equations. That is, if equilibrium equations are established for the original state of the frame, it will be assumed that the overall dimensions will alter by negligible amounts, so that the same equations can be used to describe the deformed state of the frame. It will not normally occur to the elastic designer that the equations might have to be altered to allow for deflexions. This assumption of invariance of the equilibrium equations is carried over from simple elastic design to simple plastic design.

It turns out that a large proportion of typical steel building frames has deflexions which are negligible in this sense. The arch has been mentioned as a structural type for which deflexions may be important. The pitched-roof portal frame, especially when made of high-strength steel, may also suffer rather large deflexions and care must be taken when designing such a frame by simple plastic methods. In many cases it is possible to make a simple estimate of reduction in load-carrying capacity due to deformation. If this reduction is small, say less than 5%, the designer may well decide that simple plastic theory can continue to be used with confidence. If the reduction is larger, say more than 10%, the designer may decide to abandon plastic strength calculations in favour of design to some other criterion.

It is fortunate that deflexions can be estimated fairly accurately by approximate methods. Effectively, deflexions are given by a double integration of the bending moments in a frame; if a reasonable guess can be made of the bending-moment distribution, then deflexions can be calculated to about the same degree of accuracy as those given by a conventional elastic analysis. Thus the designer can assume in the first instance that deflexions are negligible, design his frame by plastic methods, and then check that deflexions are indeed small.

Similarly, stability, for example of columns, can be checked after the plastic design has been made. The stability of an individual structural element cannot be allowed as a design criterion in conjunction with simple plastic methods. If a column in a building frame should become unstable at some stage in the loading process, this might well initiate premature collapse of the whole frame. Thus members should be proportioned on the basis of their strength in bending, and checked rigorously to see that they are stable at all stages.

Again this conclusion is not confined to structures designed by plastic theory. A well-designed elastic frame should also derive its strength from bending; a single unstable member can threaten the safety of the whole structure.

The prime requirement is, in fact, that of *ductility*. Not only should the material be ductile so that, in bending, a plastic hinge can form and suffer large rotations without drop in bending moment. The overall member behaviour must also be 'ductile' in the sense that there must be no fall-off in strength, for example due to instability.

Supposing that deflexions of a frame are small, that there is no danger of instability, and that the material properties are sufficiently good that simple plastic theory may be applied with confidence, an assessment may then be made of the fundamental aims of plastic design. Certainly the plastic designer does not proportion a structure so that it *collapses* under the *working* load; the use of the load factor ensures that there is an adequate margin of safety. In a sense, therefore, the plastic designer appears to be making purely hypothetical calculations; on the one hand, the factored collapse loads are 'unreal' and will never occur in practice, and, on the other, the designer provides no direct information as to the actual behaviour of the structure under working conditions.

In fact the *elastic* designer does not usually make a very good estimate of the working state of an actual frame, unless it happens to be statically determinate, and in most cases his estimate is so poor as to be virtually meaningless. In the two-span beam of fig. 6.1, for example, a conventional elastic calculation would indicate that the largest bending moment occurs at the central prop, and has value $Wl/8$. However, this calculation relies on the assumption that the three supports are completely rigid, or, alternatively, that if they settle they do not do so differentially. Any small differential settlement, which must inevitably occur in practice, will have a very large effect on the computed value of the bending moments. In particular, the largest bending moment, $Wl/8$, may be either reduced or increased, but it is on this single value that the whole design of a uniform beam would be based.

This simple, almost trivial, example is typical of the behaviour of structures of this kind. Structural behaviour is in practice so variable that to ask for a description of the actual state is a meaningless question. If all the variable factors, settlements, imperfect fabrication, stresses induced on erection, and so on, could be taken into account by some means, then perhaps an 'actual' bending-moment diagram could be calculated. Even such an accurate calculation would be valid only for a short time. A shift of load, for example, might cause slip in a connexion or settlement of a footing, and the original structure would then have ceased to exist, and would have been replaced by one which appears only slightly different, but whose elastic behaviour is in fact markedly different.

In the light of this uncertainty as to the actual behaviour of a frame, it seems somewhat anomalous to base the design on a computed elastic value of bending moment; moreover, it seems 'obvious' that trivial structural defects, while affecting the elastic distribution of bending moments, cannot have any real influence on the overall strength of the structure. From this point of view, plastic theory is simply a way of obtaining the maximum possible advantage from the situation.

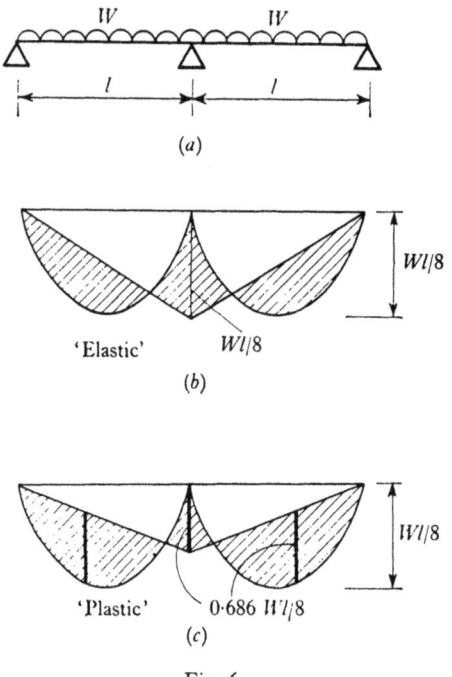

Fig. 6.1

The two bending-moment diagrams in fig. 6.1 (*b*) and (*c*) are both supposed to be due to the *working* loads. The 'elastic' distribution gives a largest bending moment of $Wl/8$, and the designer will arrange that the corresponding bending stress does not exceed a certain value, say 165 N/mm² for steel with a yield stress $\sigma_0 = 250$ N/mm². The 'plastic' designer will obtain the distribution of fig. 6.1 (*c*) from a collapse analysis, in which the loads are multiplied by a hypothetical load factor. In making his design, however, what he is really doing is to arrange that the largest *working* value of the bending moment, $0.686Wl/8$ in this case, also produces a stress which does not exceed the permitted value of 165 N/mm² (or a different permitted value if he is working to a load factor other than 1.75). In doing this, the designer will have the absolute assurance of the plastic theorems that his design cannot possibly collapse until the working loads are multiplied by a hypothetical load factor of value $250/165$ times the shape factor, that is 1.75 for the I-section.

The elastic distribution of fig. 6.1 (*b*) is believed by the elastic designer to be 'correct'; it is not. It is merely a possible equilibrium distribution of bending moments for the beam. The plastic distribution of fig. 6.1 (*c*) is another possible equilibrium distribution, and either distribution can be used, by the *plastic theorems*, as a basis for design. It is, in fact, the task of the designer to construct a 'reasonable' equilibrium state for his structure on which to base a design. For the case of steel frames, the state calculated from a conventional elastic analysis, while perhaps perfectly acceptable as a basis for design, bears almost no relation to the actual state under working loads. By contrast the plastic calculations, while also giving virtually no information about the structure under working loads, do predict very accurately the collapse load, and the designer can be confident, if he is using an adequate load factor, of the 'strength' safety of his structure under working loads.

In minimizing the largest bending moment the plastic designer will achieve economies in his structure; in fig. 6.1 (*c*) the largest bending moment has been made as small as possible, with a corresponding reduction of some 30% from the elastic figure. This equalization of moments at the direction of the designer is the key, not only to economy, but also to direct design instead of analysis. In the example of fig. 6.1 economies result because the design is based not on a *single* peak value of bending moment, but on two equalized values of bending moment. The number two occurs because the structure has a single redundancy; in a frame with R redundancies, $(R+1)$ bending moments may be equalized or, more generally if the frame is of variable section, $(R+1)$ bending stresses may be equalized.

Further, this equalization does not depend on the flexibilities of the members. Elastic design is essentially a process of trial and error, since the bending moments can only be calculated when the section properties are known, and then only with considerable labour. By contrast the plastic designer, in constructing his set of reasonable bending moments, does not have to know the section properties.

This discussion can be extended to other types of structure made from other materials. The actual state of a structure is in practice too variable for the structural forces to be calculable; so long as strength is the main design criterion, it is the task of the designer to establish a set of reasonable forces in the structure on which the design may be based. Thus although reinforced concrete can have a 'drooping' moment–curvature characteristic so that the plastic theorems cannot strictly apply, nevertheless a 'plastic' estimate of working conditions will be at least as good as an 'elastic' estimate. A plastic design can be used for reinforced concrete with as much confidence as an elastic design; it is the material that is suspect, and not the design method. In practice, reinforced concrete has proved itself to have sufficient ductility for the construction of safe structures; a reinforced concrete frame does not fail prematurely by excessive hinge rotation at critical sections.

Perhaps the most important contribution of plastic theory is that it has led to this concept of design, based on considerations of equilibrium alone, and rejecting those of compatibility and deformation. The lower bound (safe) theorem is specific: if an equilibrium state can be found which does not violate the yield condition, then however 'unlikely' that state may seem to be, the structure is safe. Moreover, the margin of safety, that is, the load factor against collapse, is predicted with great accuracy for a steel frame.

The theory applies strictly only to ductile materials like steel. However, if strength is the main design criterion, and if deflexions are small and the danger of instability negligible, then plastic theory is the fundamental tool for design of a structure made of any non-brittle material.

SECTION TABLES

Extracts from B.S. 4: Part 1: 1962 (Specification for Structural Steel Sections—Hot-rolled Sections) are reproduced by permission of the British Standards Institution, 2 Park Street, London, W. 1, from whom copies of the complete standard may be obtained.

UNIVERSAL BEAMS

| DESIGNATION | | Mass per metre | Depth of section D | Width of section B | Thickness | | Root rad. r | Depth between fillets d | Area of section | Moment of inertia | | Radius of gyration | | Elastic modulus | | Plastic modulus | |
| Serial size | Mass per foot | | | | Web t | Flange T | | | | About X-X | About Y-Y | About X-X | About Y-Y | About X-X | About Y-Y | About X-X | About Y-Y |
in	lb	kg	mm	mm	mm	mm	mm	mm	cm²	cm⁴	cm⁴	cm	cm	cm³	cm³	cm³	cm³
36 × 16½	260	387	920	420	21·5	36·6	24·1	791	493·9	717 328	42 479	38·1	9·27	15 586	2021	17 628	3 206
	230	343	911	418	19·4	32·0	24·1	791	436·9	623 868	36 250	37·8	9·11	13 691	1733	15 445	2 756
36 × 12	194	289	927	308	19·6	32·0	19·1	819	368·5	503 780	14 793	37·0	6·34	10 874	961	12 256	1 552
	170	253	918	305	17·3	27·9	19·1	819	322·5	435 802	12 512	36·8	6·23	9 490	819	10 930	1 322
	150	223	910	304	15·9	23·9	19·1	819	284·9	375 110	10 424	36·3	6·05	8 241	686	9 505	1 112
33 × 11½	152	226	851	294	16·1	26·8	17·8	756	288·4	339 130	10 662	34·3	6·08	7 971	726	9 144	1 166
	130	194	841	292	14·7	21·7	17·8	756	246·9	278 833	8 385	33·6	5·83	6 633	574	7 635	929
30 × 10½	132	196	770	268	15·6	25·4	16·5	681	250·5	239 463	7 701	30·9	5·54	6 223	575	7 156	925
	116	173	762	267	14·3	21·6	16·5	681	220·2	204 747	6 377	30·5	5·38	5 374	478	6 186	773
27 × 10	114	170	693	256	14·5	23·7	15·2	611	216·3	169 843	6 225	28·0	5·36	4 902	487	5 616	781
	102	152	688	254·5	13·2	21·0	15·2	611	193·6	150 015	5 391	27·8	5·28	4 364	424	4 989	680
	94	140	684	254	12·4	19·0	15·2	611	178·4	135 973	4 789	27·6	5·18	3 979	377	4 552	608
24 × 12	160	238	633	312	18·6	31·4	16·5	532	303·5	207 252	14 973	26·1	7·02	6 549	961	7 447	1 522
	120	179	617	307	14·1	23·6	16·5	532	227·7	151 313	10 572	25·8	6·81	4 901	689	5 512	1 092
	100	149	610	305	11·9	19·7	16·5	532	189·9	124 342	8 473	25·6	6·68	4 079	556	4 562	884
24 × 9	94	140	617	230	13·1	22·1	12·7	543	178·2	111 675	4 253	25·0	4·88	3 620	370	4 141	591
	84	125	612	229	11·9	19·6	12·7	543	159·4	98 410	3 676	24·8	4·80	3 217	321	3 672	514
	76	113	607	228	11·2	17·3	12·7	543	144·3	87 262	3 184	24·6	4·70	2 874	279	3 283	449
21 × 13	142	211	545	334	16·7	27·8	16·5	450	269·6	141 682	16 064	22·9	7·72	5 199	963	5 849	1 518
	127	189	539	332	14·9	25·0	16·5	450	241·2	125 619	14 093	22·8	7·64	4 657	850	5 212	1 340
	112	167	533	330	13·4	22·0	16·5	450	212·7	109 110	12 058	22·6	7·53	4 091	730	4 560	1 156

Size	Weight																
21 × 8¼	82	122	545	212	12·8	21·3	12·7	473	155·6	76078	3208	22·1	4·54	2794	303	3108	484
	73	109	539	211	11·6	18·8	12·7	473	138·4	66610	2755	21·9	4·46	2469	262	2820	419
	68	101	537	210	10·9	17·4	12·7	473	129·1	61531	2512	21·8	4·41	2293	239	2616	383
	62	92	533	209	10·2	15·6	12·7	473	117·6	55225	2212	21·7	4·34	2072	211	2362	340
18 × 7½	66	98	467	193	11·4	19·6	10·2	404	125·2	45653	2216	19·1	4·21	1954	230	2229	366
	60	89	464	192	10·6	17·7	10·2	404	113·8	40957	1960	19·0	4·15	1767	204	2012	325
	55	82	460	191	9·9	16·0	10·2	404	104·4	37039	1746	18·8	4·09	1610	183	1830	292
	50	74	457	190·5	9·1	14·5	10·2	404	94·9	33324	1547	18·7	4·04	1458	162	1654	260
18 × 6	55	82	465	153	10·7	18·9	10·2	404	104·4	36160	1093	18·6	3·24	1555	142	1797	229
	50	74	461	153	9·9	17·0	10·2	404	94·9	32380	963	18·5	3·18	1404	126	1620	203
	45	67	457	152	9·1	15·0	10·2	404	85·3	28522	829	18·3	3·12	1248	109	1439	176
16 × 7	50	74	413	180	9·7	16·0	10·2	357	94·9	27279	1448	17·0	3·91	1322	161	1502	257
	45	67	409	179	8·8	14·3	10·2	357	85·4	24279	1269	16·9	3·86	1186	142	1343	226
	40	60	406	178	7·8	12·8	10·2	357	75·9	21458	1104	16·8	3·81	1056	124	1191	198
	36	54	403	178	7·6	10·9	10·2	357	68·3	18576	922	16·5	3·67	923	104	1046	167
16 × 6	50	74	416	154	10·1	18·1	10·2	357	94·8	26938	1047	16·9	3·32	1294	136	1486	218
	45	67	412	153	9·4	16·0	10·2	357	85·3	23798	908	16·7	3·26	1155	119	1323	190
	40	59	408	153	8·6	13·9	10·2	357	75·8	20619	768	16·5	3·18	1011	101	1158	162
15 × 6	45	67	389	154	9·7	16·3	10·2	333	85·4	21276	947	15·8	3·33	1095	123	1254	196
	40	60	385	153	8·7	14·4	10·2	333	75·9	18632	814	15·7	3·27	968	106	1106	170
	35	52	381	152	7·8	12·4	10·2	333	66·4	16046	685	15·5	3·21	842	90	959	144
14 × 6¾	45	67	364	173	9·1	15·7	10·2	309	85·1	19483	1278	15·1	3·87	1071	148	1210	234
	38	57	359	172	8·0	13·0	10·2	309	72·1	16038	1026	14·9	3·78	894	119	1007	190
	34	51	356	171·5	7·3	11·5	10·2	309	64·5	14118	885	14·8	3·71	794	103	893	165
	30	45	352	171	6·9	9·7	10·2	309	56·9	12052	730	14·6	3·58	685	85·4	772	137
12 × 6½	36	54	311	167	7·7	13·7	8·9	263	68·3	11686	988	13·1	3·81	752	119	843	187
	31	46	307	166	6·7	11·8	8·9	263	58·8	9924	825	13·0	3·73	646	99·5	721	157
	27	40	304	165	6·1	10·2	8·9	263	51·4	8500	691	12·9	3·66	560	83·7	623	133
12 × 5	32	48	310	125	8·9	14·0	8·9	263	60·8	9485	438	12·5	2·68	611	69·9	705	112
	28	42	307	124	8·0	12·1	8·9	263	53·1	8124	367	12·4	2·63	530	59·0	609	94·9
	25	37	304	124	7·2	10·7	8·9	263	47·4	7143	316	12·3	2·58	470	51·1	539	82·3
10 × 5¾	29	43	260	147	7·3	12·7	7·6	216	55·0	6546	633	10·9	3·40	504	86·0	567	135
	25	37	256	146·4	6·4	10·9	7·6	216	47·4	5544	528	10·8	3·33	433	72·1	484	114
	21	31	251	146	6·1	8·6	7·6	216	39·9	4427	406	10·5	3·18	352	55·5	395	88·8
8 × 5¼	20	30	207	134	6·3	9·6	7·6	170	38·0	2880	354	8·71	3·05	279	52·9	313	83·7
	17	25	203	133	5·8	7·8	7·6	170	32·3	2348	280	8·53	2·95	231	41·9	259	67·0

NOTE. This table is based on universal beams with a small (2° 52') taper in the flanges. Universal beams with parallel flanges have properties at least equal to the values given.

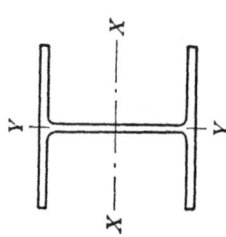

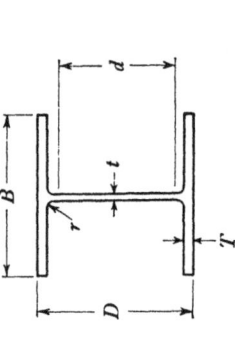

DESIGNATION			Depth of section D	Width of section B	Thickness		Root rad. r	Depth between fillets d	Area of section	Moment of inertia		Radius of gyration		Elastic modulus		Plastic modulus	
Serial size	Mass per foot	Mass per metre			Web t	Flange T				About X-X	About Y-Y	About X-X	About Y-Y	About X-X	About Y-Y	About X-X	About Y-Y
in	lb	kg	mm	mm	mm	mm	mm	mm	cm²	cm⁴	cm⁴	cm	cm	cm³	cm³	cm³	cm³
14 × 16	426	634	475	424	47·6	77·0	15·2	290	808·1	275 140	98 211	18·4	11·0	11 592	4 632	14 247	7 114
	370	550	456	418	42·0	67·5	15·2	290	701·8	227 023	82 665	18·0	10·8	9 964	3 951	12 078	6 058
	314	467	437	412	35·9	58·0	15·2	290	595·5	183 119	67 905	17·5	10·7	8 388	3 293	10 009	5 038
	264	393	419	407	30·6	49·2	15·2	290	500·9	146 765	55 410	17·1	10·5	7 004	2 723	8 229	4 157
	228	339	406	403	26·5	42·9	15·2	290	432·7	122 474	46 816	16·8	10·4	6 027	2 324	6 994	3 541
	193	287	394	399	22·6	36·5	15·2	290	366·0	99 994	38 714	16·5	10·3	5 080	1 940	5 818	2 952
	158	235	381	395	18·5	30·2	15·2	290	299·8	79 110	31 088	16·3	10·2	4 153	1 570	4 689	2 384
Column core	320	476	427	424	48·0	53·2	15·2	290	607·2	172 391	68 057	16·8	10·6	8 075	3 207	9 700	4 979

Section																	
14 × 14½	136	202	375	374	16·8	27·0	15·2	290	257·9	66307	23632	16·0	9·58	3540	1262	3976	1917
	119	177	368	372	14·5	23·8	15·2	290	225·7	57155	20470	15·9	9·53	3104	1100	3457	1668
	103	153	362	370	12·6	20·7	15·2	290	195·2	48526	17470	15·8	9·45	2681	944	2964	1430
	87	129	356	368	10·7	17·5	15·2	290	164·9	40246	14555	15·6	9·40	2264	790	2482	1196
12 × 12	190	283	365	322	26·9	44·1	15·2	247	360·4	78777	24545	14·8	8·26	4314	1525	5101	2337
	161	240	353	318	23·0	37·7	15·2	247	305·6	64177	20239	14·5	8·13	3641	1273	4245	1947
	133	198	340	314	19·2	31·4	15·2	247	252·3	50832	16230	14·2	8·03	2991	1034	3436	1576
	106	158	327	311	15·7	25·0	15·2	247	201·2	38740	12524	13·9	7·90	2368	806	2680	1228
	92	137	321	309	13·8	21·7	15·2	247	174·6	32838	10673	13·7	7·82	2049	691	2298	1052
	79	117	314	307	11·9	18·7	15·2	247	149·8	27661	9006	13·6	7·75	1756	587	1953	892
	65	97	308	305	9·9	15·4	15·2	247	123·3	22202	7268	13·4	7·67	1442	477	1589	723
10 × 10	112	167	289	265	19·2	31·7	12·7	200	212·4	29914	9796	11·9	6·78	2070	741	2417	1132
	89	133	276	261	15·6	25·3	12·7	200	168·9	22575	7519	11·6	6·68	1634	576	1875	879
	72	107	267	258	13·0	20·5	12·7	200	136·6	17510	5901	11·3	6·58	1313	457	1485	695
	60	89	260	256	10·5	17·3	12·7	200	114·0	14307	4849	11·2	6·43	1099	379	1228	575
	49	73	254	254	8·6	14·2	12·7	200	92·9	11360	3873	11·05	6·55	895	305	989	462
8 × 8	58	86	222	209	13·0	20·5	10·2	161	110·1	9462	3119	9·27	5·33	851	299	979	456
	48	71	216	206	10·3	17·3	10·2	161	91·1	7647	2536	9·17	5·28	708	246	802	374
	40	59	210	205	9·3	14·2	10·2	161	75·8	6088	2041	8·97	5·18	581	199	652	303
	35	52	206	204	8·0	12·5	10·2	161	66·4	5263	1770	8·89	5·16	510	174	568	264
	31	46	203	203	7·3	11·0	10·2	161	58·8	4564	1539	8·81	5·11	449	151	497	230
6 × 6	25	37	162	154	8·1	11·5	7·6	123·5	47·4	2218	709	6·83	3·86	274	91·8	310	140
	20	30	157	153	6·6	9·4	7·6	123·5	38·2	1742	558	6·76	3·81	221	73·1	247	111
	15·7	23	152	152	6·1	6·8	7·6	123·5	29·8	1263	403	6·50	3·68	166	52·9	184	80·9

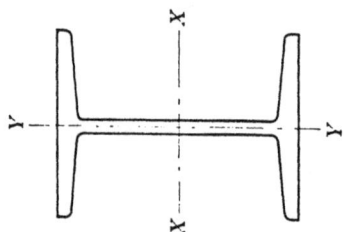

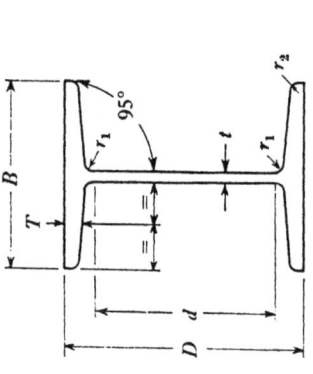

JOISTS WITH 5° TAPER FLANGES

DESIGNATION			Depth of section D	Width of section B	Thickness		Radii		Depth between fillets d	Area of section	Moment of inertia		Radius of gyration		Elastic modulus		Plastic modulus	
Serial size	Mass per foot	Mass per metre			Web t	Flange T	Root r₁	Toe r₂			About X-X	About Y-Y	About X-X	About Y-Y	About X-X	About Y-Y	About X-X	About Y-Y
in	lb	kg	mm	mm	mm	mm	mm	mm	mm	cm²	cm⁴	cm⁴	cm	cm	cm³	cm³	cm³	cm³
8 × 4	17	25·30	203·2	101·6	5·8	10·4	9·4	3·2	161·1	32·26	2294·0	162·6	8·43	2·25	225·8	32·02	256·3	51·78
7 × 4	14½	21·56	177·8	101·6	5·3	9·1	9·4	3·2	138·2	27·49	1522·0	139·7	7·44	2·25	171·2	27·50	193·4	44·61
6 × 3½	11½	17·10	152·4	88·9	4·9	8·3	7·9	2·4	117·7	21·81	883·1	86·3	6·36	1·99	115·9	19·41	131·2	31·39
5 × 3	9	13·38	127·0	76·2	4·5	7·6	7·9	2·4	94·2	17·06	477·0	50·37	5·29	1·72	75·12	13·22	85·4	21·37
4 × 2½	6¼	9·66	101·6	63·5	4·1	6·6	6·9	2·4	73·2	12·32	218·2	25·41	4·21	1·44	42·95	8·00	49·1	12·96
3 × 2	4¼	6·70	76·2	50·8	3·8	5·7	6·9	2·4	50·2	8·54	83·07	11·22	3·12	1·15	21·80	4·42	25·2	7·21

INDEX

Arch, 208, 216
Assumptions, 7
Axial load, effect of, 20; on $\underline{\text{T}}$-section, 22; on rectangular section, 20; on T-section, 26; on tubular section, 24
Axis: bending about inclined, 32; equal area, 15; neutral, 15; plastic principal, 35; zero-stress, 15, 20, shift of, 23

Beam: bending of, 6; collapse of, 1; continuous, 55; encastré, 44, 52, 124; of rectangular cross-section, 7; redundant, collapse of, 44; simply supported, 1, 36, 52
Bend test, 6
Bounds, 57, 95, 136, 220
British Standards: 4: Part I: 1962 (Specification for Structural Steel Sections—Hot rolled sections), 221; 449: The Use of Structural Steel in Building, 35, 37; 4360: Weldable Structural Steels, 2
British Standards Institution, 221

Cantilever, propped, 54
Centre, instantaneous, 99
Collapse: catastrophic, 73; deflexions, 3; equation, 123; incremental, 64; load, 3, 50; load factor, 35; mechanism, 75; mode, 71, 79, 83, 84, 99, 104, 203; partial, 60, 85; state, 51
Column, stability of, 216
Compatibility, 220; condition, 47, 51
Condition: compatibility, 47, 51; equilibrium, 132, 135, 192; mechanism, 132, 135; normality, 193; yield, 132, 135, 192
Cover plates, 17, 59
Curve: characteristic, 3; load-deflexion, 1, 6, 13, 49, 50

Deflexions, 4, 215; beam, 1, 13; cantilever, 13; unrestricted, 3
Deformation, 216, 220; of encastré beam, 46; permanent, 1
Diagram, interaction, 79, 85
Design: direct, vii, 58, 81, 219; fully rigid, 35; minimum weight, 58; of a frame, 6; of beam system, 58, 65; of cross-sections, 17; of non-uniform girder, 68; of portal, 81, 95; plane, 83; plastic, the tools of, 120; problem, 132; process, 50; simple, 35

Ductility, 217; minimum, 2

Elastic behaviour, 1, 3
Elongation, 7
Engineering, structural, vii
Engineering practice, 216
Equation, work, 68
Equilibrium, 47, 220; condition, 132; equation, 123, 144; possible state, 57
Examples, 37, 106, 141, 178, 211

Foundations, cost of, 97
Frame: complex, 57, 191; multi-storey steel, 4, 156; portal, pinned feet, 73, fixed base, 84, 124, pitched roof, 98, 128, 172; space, 191, 199

Grillage, 191

Hinge: cancellation of, 146, 154; extent of, 9; formation of, 50; plastic, 3; rotations, 148; 'rusty', 3

Imperfections, effect of, 51
Instability, 73, 215, 217
$\underline{\text{T}}$-section beam: effect of axial load, 22; in grillage, 193; plastic modulus, 12; shape factor, 13; with cover plates, 18

Load: axial, 20, in arch, 208; collapse, 3, 50; distributed, 55, 169; factor, 35, 36, 131; squash, 21, 28, of web, 31; travelling, 63; working, 35, 131, 217
Load-deflexion; characteristic, 4; curve, 6, 13, 49, 50
Loading: conditions, 1, 2; external, 1; proportional, 131

Macaulay's notation, 46
Mechanisms, 54; combination of, 120, 144, 146, 162; incomplete, 150; incorrect, 57; independent, 155; local collapse, 175; locking of, 149, 154; of collapse, 49, 53, 146; possible, 59; ridiculous, 147; scaling of, 171; true, 157; virtual, 205
Mises criterion, 29
Modulus: elastic section, 8; plastic, 6, 8; plastic, of sections with one axis of symmetry, 15

227

For EU product safety concerns, contact us at Calle de José Abascal, 56–1°,
28003 Madrid, Spain or eugpsr@cambridge.org.

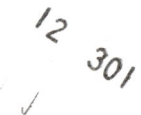

www.ingramcontent.com/pod-product-compliance
Ingram Content Group UK Ltd.
Pitfield, Milton Keynes, MK11 3LW, UK
UKHW012314141225
465965UK00001B/68